AF508671

Advanced Topics in Systems Safety and Security

Advanced Topics in Systems Safety and Security

Editors

Emil Pricop
Grigore Stamatescu
Jaouhar Fattahi

MDPI • Basel • Beijing • Wuhan • Barcelona • Belgrade • Manchester • Tokyo • Cluj • Tianjin

Editors
Emil Pricop
Petroleum-Gas University of Ploiesti
Romania

Grigore Stamatescu
University Politehnica of Bucharest
Romania

Jaouhar Fattahi
Laval University
Canada

Editorial Office
MDPI
St. Alban-Anlage 66
4052 Basel, Switzerland

This is a reprint of articles from the Special Issue published online in the open access journal *Information* (ISSN 2078-2489) (available at: https://www.mdpi.com/journal/information/special_issues/IWSSS_2019).

For citation purposes, cite each article independently as indicated on the article page online and as indicated below:

LastName, A.A.; LastName, B.B.; LastName, C.C. Article Title. *Journal Name* **Year**, *Volume Number*, Page Range.

ISBN 978-3-0365-1623-3 (Hbk)
ISBN 978-3-0365-1624-0 (PDF)

Contents

About the Editors

Emil Pricop is currently Associate Professor and Head of the Control Engineering, Computers & Electronics Department of the Petroleum-Gas University of Ploiesti, Romania. Also, he is an invited professor at the Computer Engineering Department of Faculty of Engineering, Marwadi University, India. He received his Ph.D. in Systems Engineering from Petroleum-Gas University of Ploiesti by defending the thesis "Research regarding the security of control systems" (2017). His research interest is cybersecurity, focusing mainly on industrial control systems security. Dr. Pricop is co-editor of two books published by Springer, namely Recent Advances in Systems Safety & Security (2016) and Recent Developments on Industrial Control Systems Resilience (2020). Also, he is the author or co-author of two Romanian patents, six book chapters published in books edited by Springer, and over 30 papers in journals or international conferences. From 2013, Dr. Pricop the initiator and chairman of the International Workshop on Systems Safety and Security—IWSSS, a prestigious scientific event organized annually. He is an IEEE member and has held the vice-chair position of the IEEE Young Professionals Affinity Group—Romania Section from 2017 to 2019.

Grigore Stamatescu graduated from the University Politehnica of Bucharest (UPB) in 2009 and holds the Ph.D. degree (2012) from the same university. He is currently an Associate Professor (Habil. 2019) with the Department of Automation and Industrial Informatics, Faculty of Automatic Control and Computers, UPB. His research interests include networked embedded sensing, the internet of things and distributed information processing in industry, the built environment and smart city applications. Recent results include statistical learning methods for load forecasting and anomaly detection in building energy traces and data-driven modelling of large-scale manufacturing systems, with focus on energy efficiency. Research results have been published in over 130 articles. Dr. Stamatescu was a Fulbright Visiting Scholar 2015–2016 at the University California, Merced, and a JESH Scholar of the Austrian Academy of Sciences in 2019. He is Senior Member of IEEE.

Jaouhar Fattahi is currently with Laval University in Quebec City, Canada, as an associate professor. He received his Ph.D. on cryptographic protocol security from Laval University in October 2015. He completed his postdoctoral training at the Valcartier Research Center for the Canadian Armed Forces in the field of cybersecurity. He is also a computer engineer since 1995. Jaouhar Fattahi is the author of the theory of witness functions for security verification of cryptographic protocols. He now specializes in reverse engineering and machine and deep learning applied to security and cybersecurity. He is a member of the IEEE.

Preface to "Advanced Topics in Systems Safety and Security"

The networked intelligent systems development in multiple technical areas such as industrial control systems, wireless sensor networks, automotive industry, homeland, and border protection lead to several unprecedented challenges related to their security and safety. These challenges should be addressed in a timely manner.

In this context, rapid dissemination of the research results in the field of systems safety and security is critical. That is the reason why in 2013, the first edition of the International Workshop on Systems Safety & Security (IWSSS—iwsss.org) took place in the beautiful Sinaia city, Romania. The IWSSS started with the idea of bringing together specialists and researchers interested in these topics. With every yearly edition, the workshop becomes a real platform for exchanging experience, ideas, and research results to increase awareness of cybersecurity issues and stimulate collaboration at regional and international levels.

This book is a reprint of the Information Special Issue (SI) on Advanced Topics in Systems Safety and Security. The SI comprises high-quality papers dealing with the challenging research topics in cybersecurity of computer networks and industrial control systems.

The contributions presented in this book are mainly the extended versions of selected papers presented at the 7th and the 8th editions of the International Workshop on Systems Safety and Security—IWSSS. These two editions took place in 2019 and respectively in 2020. In addition to the selected papers from IWSSS, the special issue includes other valuable and relevant contributions.

The papers included in this reprint discuss various topics ranging from cyberattack or criminal activities detection, evaluation of the attacker skills, modeling of the cyber-attacks, and mobile application security evaluation. Given this diversity of topics, we hope that the book will represent a valuable reference for researchers in the security and safety of systems.

Emil Pricop, Grigore Stamatescu, Jaouhar Fattahi
Editors

 *information*

Article

Quantitative Model of Attacks on Distribution Automation Systems Based on CVSS and Attack Trees

Erxia Li [1], Chaoqun Kang [1], Deyu Huang [2,*], Modi Hu [2], Fangyuan Chang [1], Lianjie He [1] and Xiaoyong Li [2,*]

1   China Electric Power Research Institute, Haidian District, Beijing 100192, China
2   Key Laboratory of Trustworthy Distributed Computing and Service (Beijing University of Posts and Telecommunications), Ministry of Education, Haidian District, Beijing 100876, China
*   Correspondence: huangdeyu@bupt.edu.cn (D.H.); lixiaoyong@bupt.edu.cn (X.L.)

Received: 27 May 2019; Accepted: 26 July 2019; Published: 29 July 2019

Abstract: This study focuses on the problem of attack quantification in distribution automation systems (DASs) and proposes a quantitative model of attacks based on the common vulnerability scoring system (CVSS) and attack trees (ATs) to conduct a quantitative and systematic evaluation of attacks on a DAS. In the DAS security architecture, AT nodes are traversed and used to represent the attack path. The CVSS is used to quantify the attack sequence, which is the leaf node in an AT. This paper proposes a method to calculate each attack path probability and find the maximum attack path probability in DASs based on attacker behavior. The AT model is suitable for DAS hierarchical features in architecture. The experimental results show that the proposed model can reduce the influence of subjective factors on attack quantification, improve the probability of predicting attacks on the DASs, generate attack paths, better identify attack characteristics, and determine the attack path and quantification probability. The quantitative results of the model's evaluation can find the most vulnerable component of a DAS and provide an important reference for developing targeted defensive measures in DASs.

Keywords: industrial control safety; attack quantification; common vulnerability scoring system; attack tree; distribution automation system

---

1. Introduction

1.1. Motivations

The expansion of the construction scale of distribution automation systems (DASs) and the increasing demand for their application have increased the risk of cyber and physical attacks on these systems. On 7 March 2019, Venezuela's power grid system experienced deliberate destruction [1]. Large-scale blackouts occurred in most parts of Venezuela, including its capital, Caracas, which experienced blackouts for more than 24 h. At one point, 20 of Venezuela's 23 states experienced blackouts, which seriously affected their infrastructure. In 2015, a sophisticated cyberattack targeted Ukraine's power grid and caused power outages over a wide area [2]. This highlights the importance of investment in securing power distribution grids against intruders [3]. Similarly, the overall safety of Chinese DASs must be improved, given the increasing demand for distribution network security [4]. At present, attack quantification in DASs at home and abroad remains in its infancy. DASs have high complexity and poor flexibility and lack a mature method for the quantitative evaluation of attacks on them [5–7]. Thus, ensuring DAS security has become a key challenge in the industry. To avoid disasters, defensive measures can be applied in advance through a reasonable quantitative evaluation of attacks and an evaluation of the probability of an attack on each part of a DAS [8]. Simultaneously, these attack quantification results can also provide an important reference for security technicians to implement the DAS defense system.

Quantification of the probability of an attack on a DAS directly affects the in-depth analysis of the system's security. Wang et al. [9] proposed a multilevel analysis and modeling method for a power system's communication network. Their case study showed that this method can be used to evaluate the static and dynamic relationships among power networks. Kateb et al. [10] developed an optimal structure tree method for risk assessment in a wide-area power system that can minimize the spread of network attacks. The authors in [9,10] provided a well-optimized evaluation of a specific power network. However, these evaluation neither reflected the attacker's behavior in terms of quantification of the probability of an attack nor provided suggestions for the protection of specific parts of the power system. The authors in [11] and the authors in [12] presented an attack assessment framework based on Bayes attributes—a stochastic game model and a fast modeling method for input data, respectively—which included network connection relationship and vulnerability information. However, the proposed methods were found to be inefficient when applied in DASs due to DAS architecture complexity and expansibility, and they could not generate attack path. The authors in [13] proposed a method for modeling network attacks with a multilevel-layered attack tree (MLL-AT), presented a description language based on the MLL-AT for attacks, and quantified the leaf nodes. This attack tree (AT) was found to be able to accurately model the attacks, especially multilevel network attacks, and can be used to assess system risks. However, the research is mainly based on cyberattacks, and there is no physical attacks involved. Besides, this method lacks a complete attack process identification method, and its ability to analyze attack paths is insufficient.

1.2. Main Contributions

To summarize, although a number of studies have developed measures to quantify system risks or attacks, they insufficiently describe attack behavior or attack paths. These measures are affected by subjective factors, which are unsuitable for attack quantification of distribution automation systems. To solve these problems, we propose a modeling method for quantifying attacks on DASs based on common vulnerability scoring system (CVSS) and ATs form the perspective of the attacker's behavior. The proposed node attack probability quantification algorithm combined with the CVSS has favorable expansibility. This algorithm can improve the probability of predicting attacks on DASs, generate attack paths, and discover the latest protection component.

To our knowledge, this study is the first to use the AT to quantify the probability of attacks in DASs, which is systematic and quantitative evaluation of attacks in DASs. The main contributions are as follows.

- First, a DAS security architecture is developed on the basis of the functional characteristics and security protection requirements of DASs. This architecture provides an intuitive view of the security components of a DAS, which can help system designers have a clear understanding of the path to possible cyber-attacks and physical-attacks.
- Second, a DAS attack quantification model was established by forming a set of complete attack processes and paths based on attacker behavior, which can help DAS security practitioners to find the system components that should be defended, helping penetration testers to deploy targeted and focused attacks.
- Third, a quantification algorithm for attack probability based on an AT and the CVSS was proposed. This algorithm reduces the influence of the subjective factors in the process for quantifying attacks in traditional approaches and improves the accuracy of attack prediction. The efficacy of the model was evaluated by introducing the environmental characteristics of the DAS.

The experimental results show that the proposed model can predict the risk of attack that the DAS faces. The results of the model's evaluation verify feasibility, effectiveness of the proposed scheme and provide an important reference for the development of targeted defensive measures for DASs.

The rest of this paper is structured as follows: Section 2 gives a detailed design of the DAS security architecture. The quantitative model of attacks on DAS based on CVSS and ATs is described in Section 3. The experimental results are presented in Section 4. Finally, Section 5 concludes the paper.

2. Design of the DAS Security Architecture

DASs have the characteristics of a large number of terminals, high complexity architecture, poor flexibility, and require strict protection against both network attacks and physical attacks [14]. A DAS security architecture was developed on the basis of the functional characteristics and security protection requirements of DASs. It is shown in Figure 1.

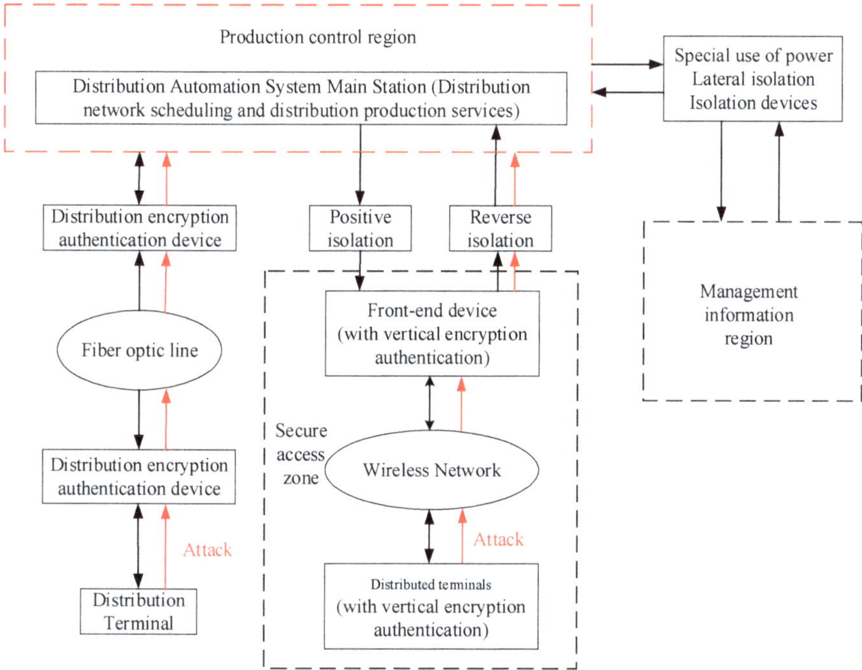

Figure 1. Distribution automation system (DAS) security architecture.

(1) The production control region directly manages the distribution automation system's main station and controls the automatic power distribution scheduling of the entire distribution network. It is at the core of the DAS's distribution scheduling and production services. It includes the main station's server, the main station's monitoring computing station, the main station's transport unit controller, and other equipment, which are vulnerable to phishing, distributed denial-of-service attacks, and physical attacks [15].

(2) The communication mode of the application part of the management information region is based mainly on public network communication. It is connected to the production control region by an isolation device to realize a large amount of data storage and thus is very sensitive to Web data security.

(3) The secure access zone includes wireless network, some acquisition servers, and the front-end device that transmits commands and collects terminal data so that the DAS can realize intelligent power distribution and optimized operation. As the link between the core of the distribution network and the terminal information exchange, this zone faces many security risks. An attacker can use the terminal as a springboard to invade or attack through the wireless network.

(4) At the furthermost edge of the DAS is the power distribution terminal. It can communicate with the main station through an optical fiber. Although this part of the equipment is a great distance away from the core equipment for power distribution, it is the smallest unit and supplies power to the distribution automation system. It is the point of the system that is most vulnerable to attacks.

3. DAS Attack Quantification Algorithm

In order to face the different security attacks that can occur in the DAS security architecture, an attack probability quantification model based on an AT for the DAS framework is proposed. Each leaf node of the AT represents an attack on a certain component of the DAS security architecture. The maximum probability of each attack path in ATs will be calculated on the basis of the CVSS in terms of three measurement factors— base, time, and environment.

3.1. DAS AT Model

The AT was first proposed by Schneier [16]. In the structure of an AT, the root node represents the target of the attack [17]. The characteristics of system security are described on the basis of the AT. These descriptions redefine the data on attacks by identifying whether the DAS security or survival criteria are satisfied, and the data are regarded as the root nodes of the tree. In Figure 2, a node represents the means of implementing an attack, and the relationship among the nodes may be the logical OR, that is, the attack target can be reached when one of the two nodes E1 and E2 satisfies the attack conditions; AND, that is, the attack target can be reached when nodes E1 and E2 satisfy the attack conditions simultaneously; or Order AND, that is, when the attack target is reached after nodes E1 and E2 satisfy the attack conditions [18]. The AT has the advantages of simple structure, easy to understand presentation method, and easy to focus the analysis process on measurable targets. It can be combined with the obvious features of DAS in terms of architecture and simplify the DASs of system security features.

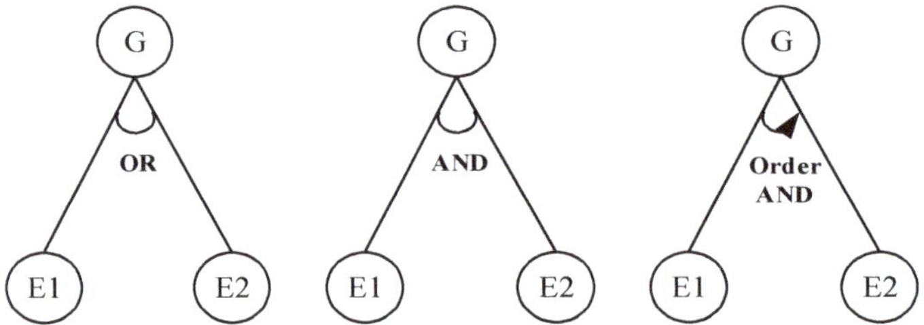

Figure 2. Node representations in the attack tree (AT).

The DAS AT model must consider the environment and the DAS security architecture. Figure 3 shows the main stages of the DAS AT model. The nodes of all leaves will first be quantified when the ATs are established. Then, the probability of a successful attack in all paths of the system will be calculated by modeling the DAS AT. The attack path sequence is obtained through calculation, and the path with the maximum attack probability is the optimal attack path.

The use of software vulnerabilities is a well-known way to attack a network. Our attack probability quantification algorithm is based on the CVSS. The attack probability value of the Common Vulnerabilities and Exposures (CVE) vulnerabilities at each node of the DAS is calculated using the CVSS method. Furthermore, combined with the method of attacking the tree, each path the attack probability of the DAS is calculated to evaluate the probability of each attack.

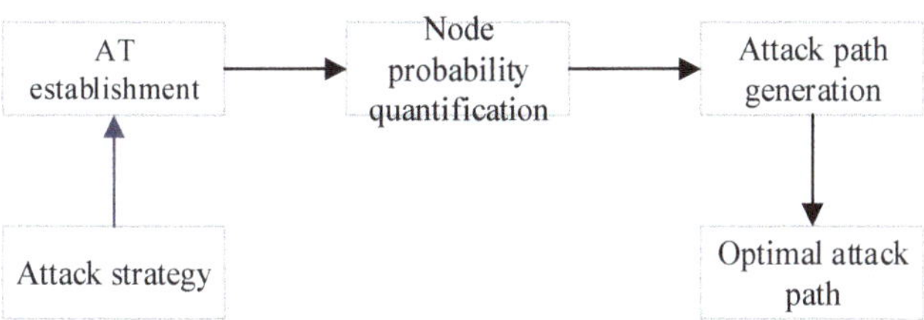

Figure 3. The main stages of the DAS AT Model.

3.2. CVSS

The CVSS is a standard for calculating the risk level of each CVE vulnerability. It was developed by the National Infrastructure American Council and is maintained by the Forum of Incident Response and Security Teams [19]. Manufacturers can adopt this system for free. On the basis of the CVSS, we can score a system's weaknesses and determine which weaknesses have priority for repair. The CVSS provides an open framework for evaluating the characteristics and impact of system vulnerabilities for information security industry–related practitioners. The CVSS quantifies CVE vulnerabilities using scores (0–10) of severity, and strict attack indexes can be formulated, including attack vector, attack complexity, authentication, availability, integrity, and confidentiality indexes [20].

As shown in Figure 4, the CVSS consists of three basic score indicators, namely the base score, the temporal score, and the environmental score. The base score includes exploitability metrics and impact metrics, which have their own calculation formulas. The temporal and environmental scores can be expanded. Moreover, a vector string and a CVSS score, which represent the calculation process and the result, respectively, are generated.

The CVSS is supported by the National Vulnerability Database (NVD) of the United States. All CVE vulnerabilities in the NVD contain the basic value of the CVSS [21]. The quantification of the DAS attack probability is closely related to the evaluation indexes of vulnerabilities for all parts of a DAS and plays an important auxiliary role in the quantification of an attack process in the DAS. Thus, the probability of attack that the DAS faces is quantified on the basis of the CVSS.

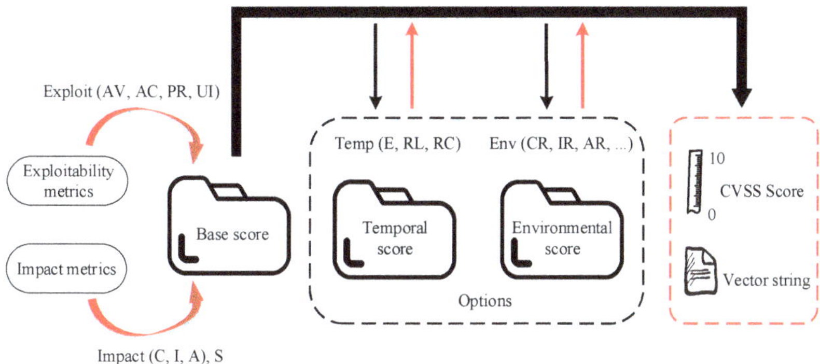

Figure 4. Score calculation in the common vulnerability scoring system (CVSS).

Table 1 lists the relevant variables for calculating the CVSS base score [22]. In accordance with these variables, the base score represents the inherent characteristics of the vulnerability itself and the possible impact of these characteristics. The scoring situation can determine the attack probability that the vulnerability represents.

Table 1. Base score calculation-related metrics.

Relevant Metrics	Possible Metric Values	Quantified Scores
Attack vector (AV)	Network (N)/Adjacent (A)/Local (L)/Physical (P)	0.85/0.62/0.55/0.2
Attack complexity (AC)	Low (L)/High (H)	0.77/0.44
Privilege required (PR)	Non (N)/Low (L)/High (H)	0.85/0.62/0.27
User interaction (UI)	Non (N)/Requirement (R)	0.85/0.62
Scope of influence (S)	Unchanged (U)/Changed (C)	Depends on ESS, ISC
Confidentiality (C)	Non (N)/Low (L)/High (H)	0/0.22/0.56
Integrity (I)	Non (N)/Low (L)/High (H)	0/0.22/0.56
Availability (A)	Non (N)/Low (L)/High (H)	0/0.22/0.56

For example, the scoring rubric for Attack Vector (AV) is divided into four possible metric methods. Figure 5 shows the division of measurement methods [22]. The score increases in the direction of the arrow in the figure. For example, the metrics of Network (N) and Adjacent (A) are the vulnerable components via the network stack, and the metrics of Local (L) and Physical (P) require physical access to the target. Network (N) can be exploited from across a routed network, which makes it easier to implement network attacks, so the measurement value is higher. However, the metric of Adjacent (A) is only exploitable across a limited logical or physical network distance.

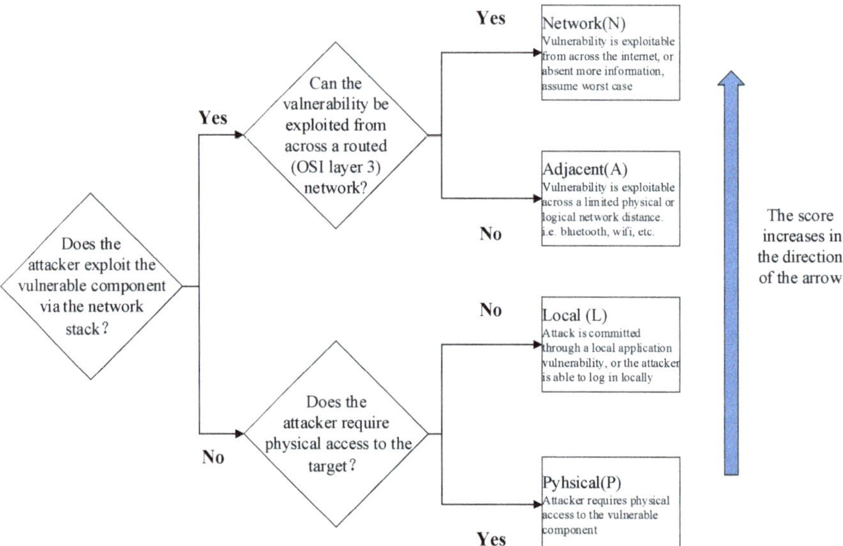

Figure 5. The scoring rubric for the Attack Vector metric.

3.3. Attack Probability Quantification Algorithm Based on the AT

To quantify the attack probability of the entire DAS, we must first determine the attack probability of each key module (leaf node) in the DAS. Second, all of the potential attack paths in the AT need to be traversed to count the probability of each path and determine the most probable attack path. On the basis of the CVSS characteristics, the vulnerability attack probability P_{attack} of a leaf node is defined as

$$\frac{Base\ Score + Temp\ Score + Envi\ Score}{10.0 * (1 + n)}, \tag{1}$$

where *Temp Score* and *Envi Score* denote the temporal and environmental scores, respectively, which can be expanded by a vulnerability to a user's environment. However, base score is a mandatory option, but scoring the Temporal and Environmental metrics is optional. n denotes the number of temporal scores and environmental scores. The *Base Score* consists of the exploitability sub score (ESS) and the impact sub score (ISC). ESS and ISC are related to the scope of influence in the factors (scope). The Base Score value is calculated using Algorithm 1 [22].

Algorithm 1. Calculate the value of the Base Score.

Input: ESS (Exploitability Sub Score); ISC (Impact Sub Score)
Output: Base Score
(1) **procedure**: *Base Score, Roundup*()
(2)　　**If** ISC $<=$ 0 **then**
(3)　　　　*Base Score* = 0
(4)　　**else if** ScopeUnchanged **then**
(5)　　　　*Base Score = Roundup* (*Minimum* [(ESS + ISC), 10])
(6)　　**else**
(7)　　　　*Base Score = Roundup* (*Minimum* [1.08 × (ESS + ISC), 10])
(8)　　**end if**
(9) **end procedure**

The ISC, which is determined by the confidentiality, integrity, and availability indexes, is calculated using Algorithm 2.

Algorithm 2. Calculate the value of impact sub score (ISC).

Input: *ImpactConf*; *ImpactInteg*; ImpactAvail
Output: ISC
(1) **procedure**: ISC
(2)　　ISCtmp = 1 − [(1 − ImpactConf) × (1 − ImpactInteg) × (1 − ImpactAvail)]
(3)　　**if** ScopeUnchanged **then**
(4)　　　　ISC = 6.42 × *ISCtmp*
(5)　　**else if** Scopechanged **then**
(6)　　　　ISC = 7.52 × [*ISCtmp* − 0.029] − 3.25 × [*ISCtmp* − 0.02]^{15}
(7)　　**end if**
(8) **end procedure**

The relationships between ESS and Attack Vector (AV), ESS and Attack Complexity (AC), ESS and Privileges Required (PR), and ESS and user interaction (UI) are expressed as

$$ESS = 8.22 \times AV \times AC \times PR \times UI. \tag{2}$$

After calculating the attack probability of a single node, the formula for calculating the probability of a successful attack at the parent node is based on two nodes, namely the AND and OR nodes.

(1)　For the AND or Order AND node, the attack probability of the current parent node G is the product of the attack probability at the child nodes.

$$P_{\text{attack}}(G) = \prod_{i=1}^{n} P_{\text{attack}}(Gi) \tag{3}$$

(2)　For the OR node, the attack probability of the parent node G is the maximum attack probability of the child nodes.

$$P_{\text{attack}}(G) = \max\{P(G1), P(G2), \ldots, P(Gn)\} \tag{4}$$

A traversal from a leaf node to a root node represents a possible attack path within the DAS. Based on the calculation of the attack probability at the AND and OR nodes, the target node that attacks a certain attack path $S_j = \{G_i|\, i = 1, 2, \ldots, n\}$ is set as G, and the probability of a successful attack is

$$P_{\text{attack}}(S_j) = \prod_{i=1}^{n} P_{\text{attack}}(Gi) \tag{5}$$

When $P_{\text{attack}}(S_j)$ is high, both the probability of a successful attack and the risk factor of the system will also be high. Thus, a defense can be firmly mounted. The maximum attack probability of the entire system can be expressed as

$$P_{\text{attackmax}}(S) = \max\{P(S_1), P(S_2), \ldots, P(S_j)\} \tag{6}$$

4. Experimental Evaluation

To verify the feasibility and effectiveness of the attack probability quantification algorithm, an attacker model was established through the quantification algorithm, and an experimental environment was built. The comparison was performed using a quantification algorithm from the literature.

4.1. Construction of the Experimental Environment

An attacker's abilities, state, and DAS-related information should be determined before quantitative modeling. These data are used as a bridge between the attacker behavior and a system attack probability analysis. An attacker can launch an attack from anywhere inside or outside the system. On the basis of an attacker's worst possible attack behavior [23,24], we adopt the following assumptions: (1) attackers are knowledgeable about the DAS and have up-to-date DAS vulnerability information, (2) attackers can deliberately and effectively attack using social engineering, (3) the minimum expected attack income gains are obtained before an attacker attacks, and (4) effective attacks frequently have a few atomic attack steps.

In this group of experiments, the AT is built to destroy the safe operation of the DAS. The DAS AT and attack paths were established as shown in Figure 6 on the basis of attackers' behavior and all the vulnerability and possible attacks of various components of the actual system in Section 2. Each leaf node of the AT represents an attack on a certain component of the DAS security architecture. After the leaf node attack probability has been calculated, the leaf node that is set back from a leaf node traversal to the root node generates a complete attack path. A root node indicates that the attack has reached G. On the basis of the different types of attacks, intrusions into the DAS can be divided into G1 (a network attack through the distribution terminals and the information management region) and G2 (an attack through the physical equipment in the production control region). The system is captured and loss is caused when any attack on G1 and G2 occurs.

Table 2 presents the definitions for all nodes in the DAS AT shown in Figure 5 together with the DAS security architecture. For example, in the attack path E5 > H3 > H1 > G1, H3 denotes an attack after acquiring a puppet machine and is an OR node, which requires one of the leaf nodes to be attacked (e.g., E2, E3, E4, or E5). After a remote network attack, E5 implants a virus-controlled puppet (H3), thereby making it reach G1 through an Internet attack (H1) and invade G to achieve a complete attack. Path E6 > H2 > G1 > G indicates that leaf node E6 reaches G1 through H2 (an internal local area network (LAN)) to crack the internal wireless network password and obtain traffic information, thus breaking into the DAS to achieve intrusion.

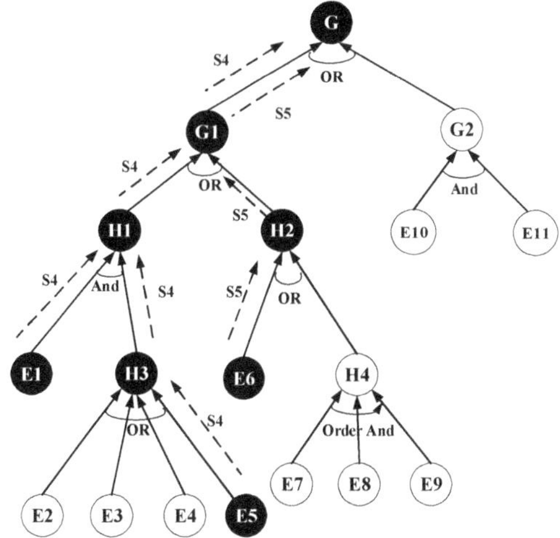

Figure 6. The DAS AT and attack path.

Table 2. Definitions for all nodes in the DAS AT.

Nodes	Definitions
G	Damaging/intruding into the DAS, endangering security
G1	Reaching G through a network attack
G2	Reaching G through an attack on physical equipment
H1	Reaching G1 through an Internet attack
H2	Reaching G1 through internal and related business network attacks
H3	Attack after acquiring a puppet machine
H4	Acquiring sensitive information from the internal database
E1	Implanting a Trojan horse into the control server
E2	Obtaining server data through phishing mail/web pages
E3	Intruding through a distributed denial-of-service attack
E4	Obtaining data by invading a web service of the DAS on the Internet
E5	Intruding through remote network vulnerabilities
E6	Cracking an internal wireless network password to obtain traffic information
E7	Scanning internal network port, service, and other asset information
E8	Acquiring root access to the database
E9	Attempting remote code execution through SMB vulnerabilities
E10	Entering into the distribution automation system through social engineering
E11	Breaking the BIOS through a u-disk to bypass a password requirement

4.2. Analysis of the Experimental Results

The CVE vulnerability numbers were established on the basis of the attack characteristics of each leaf node $\{E_i|\ i = 1, 2, \ldots, n\}$ and a DAS enterprise vulnerability evaluation in order to reflect the generality of the system components in the experiment involving a DAS while avoiding an attack-oriented experiment involving a hacker. System component vulnerabilities are not fully exploitable vulnerabilities in current DASs but rather represent vulnerabilities with different vendor components of the same type. For example, we choose the vulnerability number CVE-2018-0247 that is same type of vulnerability of Cisco Wireless LAN Controller for E6. The vulnerability attack probability of each leaf node was calculated by combining Equations (2) and (3) with Algorithms 1 and 2. Table 3 summarizes the DAS components and the vector string and P_{attack} results.

For example, E1 denotes the embedding of a Trojan horse into the control server. This activity occurs in the distribution encryption authentication device. The corresponding vulnerability number is CVE-2017-5873, and its vector string is AV:L/AC:L/PR:H/UI:N/S:U/C:H/I:H/A:H. AV is local (0.55), AC is low (0.77), PR is high (0.27), UI is unnecessary (0.85), the scope of influence (S) is unchanged, confidentiality (C) is high (0.56), integrity (I) is high (0.56), and availability (A) is high (0.56). Based on Algorithm 2, in the case of *ScopeUnchanged*, *ISCtmp* is correlated with AC, AV, and PR; that is, $1 - (1 - 0.55) \times (1 - 0.77) \times (1 - 0.27) = 0.924445$, and the ISC is $6.42 \times ISCtmp = 5.9$. Based on Equation (3), the ESS is $8.22 \times AV \times AC \times PR \times UI = 0.8$. Furthermore, in combination with Algorithm 1, the *Base Score* is 6.7. Based on Equation (2), P_{attack} is 0.67. Table 3 shows that the probability of a successful attack on the distributed encryption authentication device at this node using its vulnerability number (CVE-2017-1287) is more than 60%.

As shown in Table 4, the AT contains seven attack paths, namely S1 = (E1, E2), S2 = (E1, E3), S3 = (E1, E4), S4 = (E1, E5), S5 = (E6), S6 = (E7, E8, E9), and S7 = (E10, E11).

Table 3. Results on the attack probability of the DAS nodes.

Leaf Nodes	Vulnerability No.	DAS Components	Vector String	P_{attack}
Figure 6E1	CVE-2017-1287	Distributed encryption authentication device	AV:N/AC:L/PR:L/UI:R/S:C/C:L/I:L/A:N	0.67
E2	CVE-2017-5873	Management information region terminal	AV:L/AC:L/PR:H/UI:N/S:U/C:H/I:H/A:H	0.54
E3	CVE-2018-1137	Front-end device	AV:N/AC:L/PR:L/UI:N/S:U/C:N/I:H/A:H	0.81
E4	CVE-2017-5873	Management information region terminals	AV:N/AC:L/PR:H/UI:R/S:C/C:L/I:L/A:N	0.48
E5	CVE-2018-9935	Distributed terminal	AV:N/AC:L/PR:N/UI:R/S:U/C:H/I:H/A:H	0.88
E6	CVE-2018-0247	Wireless network in the security access region	AV:A/AC:L/PR:N/UI:N/S:C/C:N/I:L/A:N	0.47
E7	CVE-2015-6314	Production control region server	AV:N/AC:L/PR:N/UI:N/S:U/C:H/I:H/A:H	0.98
E8	CVE-2015-596	Acquisition server in the security access region	AV:L/AC:L/PR:N/UI:N/S:U/C:H/I:N/A:N	0.62
E9	CVE-2018-3269	Production control region server	AV:N/AC:L/PR:L/UI:N/S:U/C:N/I:N/A:L	0.53
E10	CVE-2017-2839	Monitoring computing station in the main station	AV:N/AC:H/PR:N/UI:N/S:U/C:N/I:N/A:H	0.59
E11	CVE-2009-0243	Transport unit controller in the main station	AV:L/AC:L/UI:N/C:C/I:C/A:C	0.72

On the basis of the leaf node attack probability (Table 3), the probability on each attack path can be calculated by combining Equations (4)–(7). Each serial number represents an attack path sequence. Table 4 displays the results of the calculation of the attack path sequence probabilities. S6 represents E7, E8, and E9. In Figure 2, the node is Order AND. Based on Equation (4), $P_{attack}(H4)$ is the product of E7, E8, and E9; that is, $0.98 \times 0.62 \times 0.53 = 0.322$. Based on Equation (6), the maximum probability of $P_{attack}(G)$ is the attack probability of S4; that is, 0.5896. These results show that the maximum probability of a successful attack on the existing DAS is greater than 50%.

Table 4. Results of the calculation of the attack path probability.

Serial No.	Attack Paths	Attack Probability
S1	E1, E2, G1, G	0.3618
S2	E1, E3, G1, G	0.5427
S3	E1, E4, G1, G	0.3216
S4	E1, E5, G1, G	0.5896
S5	E6, G1, G	0.47
S6	E7, E8, E9, G1, G	0.322
S7	E10, E11, G2, G	0.4248

The Bayes method [11] was compared with the proposed attack probability quantification method to verify the latter's accuracy. The Bayes method aims to quantitatively evaluate the vulnerability of computer networks using a Bayes attribute attack graph and the CVSS. Figure 7 shows the results of the comparison. The two methods for evaluating attack sequence probability exhibit different performance with respect to highlighting risky paths. Figure 7 shows that the proposed AT model obtains a higher attack probability than the Bayes method when evaluating paths S2 and S4. In an actual DAS architecture displayed in Figure 1, the attack probability on paths S2 and S4 is the highest, which represents E1 Distributed encryption, E3 Front-end device, and E5 Distributed terminal in DASs. The probability result of the attack sequences obtained by the two methods are slightly different, and both S2 and S4 are the attack paths with the highest risk probability, which also verifies the reliability and validity of the proposed method. On the other hand, the proposed AT model probability is higher than the Bayes method probability 4.02%—6.11% in conditions S2 and S4. From Figure 7, the proposed method probability of attack is higher than the Bayes method, and the experimental result is conducive to security practitioner to pay more attention to the protection of dangerous parts of DASs.

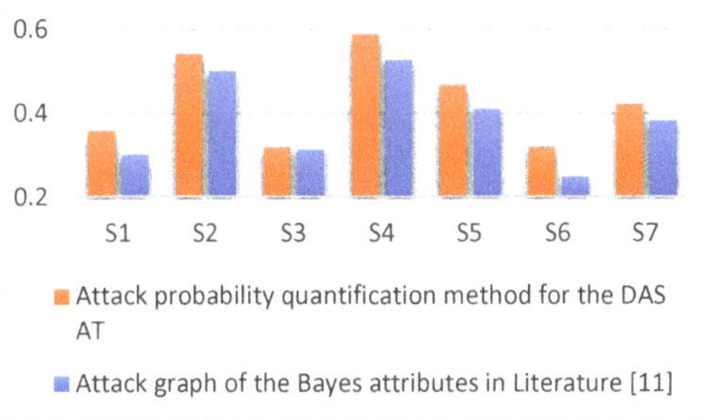

Figure 7. Comparison of the attack path probability for the DAS cases.

Due to the adoption of AT to construct DAS security architecture and attack paths, the advantage of this method is that it has more accurate probability calculation ability for network attacks and also more suitable for DASs with complex hierarchical network structure. The DAS attack quantification model is established by forming a set of complete attack processes and paths based on attacker behavior, which can help DAS security practitioners to find the system components that should be defended and help penetration testers to deploy targeted and focused attacks.

Compared with the Bayes method, the AT has the advantages of simple structure, and it is easy to focus the analysis process on measurable targets. It can be combined with the obvious features of DAS in terms of architecture and simplify the DASs of system security features. The logical "OR" and the logical "AND" characteristics of AT are very beneficial to construct such a complex DAS. At the same time, combining the characteristics of AT and DASs based on attacker behavior generated all the attack paths. Taken together, the proposed method is more effective than the Bayes method.

This finding reflects that an attack will succeed if the attackers have an abundance of information on the system. When combined with the actual security situation of the DAS, the experimental result predicts the danger of these paths and helps us to determine the components that must be defended considering that these components provide the DAS with effective defense solutions. Therefore, the proposed method is more effective than the Bayes method.

Figure 8 shows the proportions of all attack paths for the DAS. The DAS attack risks of each attack path in the system are emphasized, and the most dangerous part of the system is identified. Table 4 and Figure 8 show that the most profitable attack sequences for attackers are S2 and S4 in this experiment, and the corresponding attack methods are distributed denial-of-service attacks and website intrusions. Therefore, the DAS security practitioners should spend more time focusing on defending against these associated attacks and system vulnerabilities. For example, defense measures for the network traffic at the web end and the main station's server could be applied.

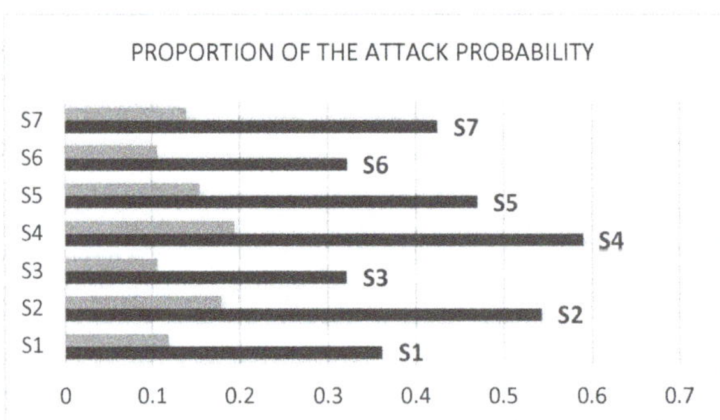

Figure 8. Proportions of the attack path probability for the DAS cases.

The evaluation methods [9–13] are based on a vulnerability analysis of traditional computer nodes and cannot quantify the attack probability of DASs. The proposed attack probability quantification algorithm and attack path calculation method can describe the vulnerability of the target system component of the DAS. To improve the accuracy of the quantification based on Algorithms 1 and 2, a set of complete attack processes and paths was constructed. The attack path with the maximum probability (Table 4) was determined to help security personnel find the attack path and DAS components with the most defense.

5. Conclusions

DASs are important to national infrastructures, which have experienced increasingly serious threats to information security. The safe and reliable operation of a DAS is directly related to the national economy and people's livelihood. In this study, a quantitative and systematic evaluation of DAS attacks was performed by analyzing the literature on attack quantification and the characteristics of the DAS environment. A modeling method for quantifying DAS attacks based on the CVSS and an AT was presented, and its feasibility was verified through experiments.

To our best knowledge, this work is the first to quantify attack value by ATs in DASs. The AT model is very suitable for DASs hierarchical features in architecture. The experimental results show that the proposed model can reduce the influence of subjective factors on attack quantification, improve the probability of predicting attacks on the DASs, generate attack paths, better characterize attack characteristics, and determine the attack path and quantification probability. The quantitative results of the model's evaluation can find the most vulnerable component of a DAS and provide an important reference for developing targeted defensive measures in DASs.

Author Contributions: Conceptualization, E.L. and C.K.; methodology, D.H., M.H., and X.L.; validation, F.C., L.H., D.H., and M.H.; formal analysis, E.L. and D.H.; investigation, C.K. and D.H.; resources, E.L. and X.L.; writing—original draft preparation, D.H.; writing—review and editing, E.L., X.L., and C.K.; supervision, X.L.; project administration, E.L. and X.L.; funding acquisition, E.L. and C.K.

Funding: This research was supported in part by the project "Research on Security Architecture for Next Generation Distribution Automation System" of the State Grid Corporation of China and was partly supported by the National Nature Science Foundation of China (61672111).

Acknowledgments: The authors would like to convey their heartfelt gratefulness to the reviewers and the editor for their valuable suggestions and important comments that greatly helped to improve the presentation of this manuscript.

Conflicts of Interest: The authors declare no conflicts of interest.

References

1. Dobson, P.; Koerner, L.; Vaz, R. Venezuela: Guaido stripped of immunity, protests erupt over blackouts. *Green Left Weekly* **2019**, *1216*, 13.

2. Zhou, L.; Ouyang, X.; Ying, H.; Han, L.; Cheng, Y.; Zhang, T. Cyber-Attack Classification in Smart Grid via Deep Neural Network. In Proceedings of the 2nd International Conference on Computer Science and Application Engineering, Hohhot, China, 22–24 October 2018.

3. Sun, C.C.; Hahn, A.; Liu, C.C. Cyber security of a power grid: State-of-the-art. *Int. J. Electr. Power Energy Syst.* **2018**, *99*, 45–56. [CrossRef]

4. Zhao, Y.; Bai, M.; Liang, Y.; Ma, J.; Deng, P. Fault Modeling and Simulation Based on Cyber Physical System in Complex Distribution Network. In Proceedings of the 2018 China International Conference on Electricity Distribution (CICED), Tianjin, China, 17–19 September 2018; pp. 1566–1571.

5. Ciapessoni, E.; Cirio, D.; Massucco, S.; Morini, A.; Pitto, A.; Silvestro, F. Risk-based dynamic security assessment for power system operation and operational planning. *Energies* **2017**, *10*, 475. [CrossRef]

6. Huang, K.; Zhou, C.; Qin, Y.; Tu, W. A Game-Theoretic Approach to Cross-Layer Security Decision-Making in Industrial Cyber-Physical Systems. *IEEE Trans. Ind. Electron.* **2019**. [CrossRef]

7. Huang, W.; Liu, Q.; Yang, S.; Xiong, W.; Liu, Z. Security situation awareness based on power-supply ability model of active distribution system. *Electr. Power Autom. Equip.* **2017**, *37*, 74–80.

8. Hahn, A.; Ashok, A.; Sridhar, S.; Govindarasu, M. Cyber-physical security testbeds: Architecture, application, and evaluation for smart grid. *IEEE Trans. Smart Grid* **2013**, *4*, 847–855. [CrossRef]

9. Wang, Q.; Pipattanasomporn, M.; Kuzlu, M.; Tang, Y.; Li, Y.; Rahman, S. Framework for vulnerability assessment of communication systems for electric power grids. *IET Gener. Transm. Distrib.* **2016**, *10*, 477–486. [CrossRef]

10. Kateb, R.; Tushar, M.H.K.; Assi, C.; Debbabi, M. Optimal tree construction model for cyber-attacks to wide area measurement systems. *IEEE Trans. Smart Grid* **2018**, *9*, 25–34. [CrossRef]

11. Wang, X.; Sun, B.; Liao, Y.; Xiang, C. Computer Network Vulnerability Assessment Based on Bayesian Attribute Network. *J. Beijing Univ. Posts Telecommun.* **2015**, *38*, 106–112.

12. Miao, F.; Zhu, Q.; Pajic, M.; Pappas, G.J. A hybrid stochastic game for secure control of cyber-physical systems. *Automatica* **2018**, *93*, 55–63. [CrossRef]

13. Yan, F.; Yin, X.; Huang, H. Research on establishing network intrusion modeling based on MLL-AT. *J. Commun.* **2011**, *32*, 115–124.

14. Zhang, H.; Wu, Z.; Ge, F.; Rong, X.; Yang, W.; Xu, C. Research on Power Distribution Automation Construction Effects Evaluation System Based on SMART Criteria. *Power Syst. Technol.* **2016**, *40*, 2192–2198.

15. Luo, F.; Yang, W.; Zhang, T.; Wang, C.; Wei, G.; Yao, L. Influence of Distribution Automation Data Transmission Errors on Power Supply Reliability of Distribution System. *Autom. Electr. Power Syst.* **2018**, *42*, 10–19.

16. Schneier, B. Attack trees. *Dr. Dobb's J.* **1999**, *24*, 21–29.

17. Lallie, H.S.; Debattista, K.; Bal, J. An empirical evaluation of the effectiveness of attack graphs and fault trees in cyber-attack perception. *IEEE Trans. Inf. Forensics Secur.* **2018**, *13*, 1110–1122. [CrossRef]

18. Kong, H.K.; Hong, M.K.; Kim, T.S. Security risk assessment framework for smart car using the attack tree analysis. *J. Ambient Intell. Hum. Comput.* **2018**, *9*, 531–551. [CrossRef]

19. Doynikova, E.; Kotenko, I. CVSS-based probabilistic risk assessment for cyber situational awareness and countermeasure selection. In Proceedings of the 2017 25th Euromicro International Conference on Parallel, Distributed and Network-Based Processing (PDP), St. Petersburg, Russia, 6–8 March 2017; pp. 346–353.

20. Venkataramanan, V.; Srivastava, A.; Hahn, A.; Zonouz, S. Enhancing Microgrid Resiliency Against Cyber Vulnerabilities. In Proceedings of the 2018 IEEE Industry Applications Society Annual Meeting (IAS), Portland, OR, USA, 23–27 September 2018; pp. 1–8.

21. Aksu, M.U.; Dilek, M.H.; Tatlı, E.İ.; Bicakci, K.; Dirik, H.İ.; Demirezen, M.U.; Aykır, T. A quantitative CVSS-based cyber security risk assessment methodology for IT systems. In Proceedings of the 2017 International Carnahan Conference on Security Technology (ICCST), Madrid, Spain, 23–26 October 2017; pp. 1–8.

22. Common Vulnerability Scoring System v3.0: User Guide. Available online: https://www.first.org/cvss/v3.0/user-guide (accessed on 19 June 2019).

23. Moussa, B.; Debbabi, M.; Assi, C. A detection and mitigation model for PTP delay attack in an IEC 61850 substation. *IEEE Trans. Smart Grid* **2018**, *9*, 3954–3965. [CrossRef]

24. Li, X.; Ma, H.; Zhou, F.; Gui, X. Service operator-aware trust scheme for resource matchmaking across multiple clouds. *IEEE Trans. Parallel Distrib. Syst.* **2015**, *26*, 1419–1429. [CrossRef]

 *information*

Article

Early-Stage Detection of Cyber Attacks

Martina Pivarníková [1], Pavol Sokol [2,*] and Tomáš Bajtoš [2]

[1] Ministry of Investments, Regional Development and Informatization of the Slovak Republic, Computer Security Incident Response Team Slovakia - CSIRT.SK, 811 05 Bratislava, Slovakia; martina.pivarnikova@csirt.sk

[2] Faculty of Science, Pavol Jozef Šafárik University in Košice, 040 01 Košice, Slovakia; tomas.bajtos@upjs.sk

*   Correspondence: pavol.sokol@upjs.sk

Received: 1 October 2020; Accepted: 26 November 2020; Published: 29 November 2020

Abstract: Nowadays, systems around the world face many cyber attacks every day. These attacks consist of numerous steps that may occur over an extended period of time. We can learn from them and use this knowledge to create tools to predict and prevent the attacks. In this paper, we introduce a way to sort cyber attacks in stages, which can help with the detection of each stage of cyber attacks. In this way, we can detect the earlier stages of the attack. We propose a solution using Bayesian network algorithms to predict how the attacks proceed. We can use this information for more effective defense against cyber threats.

Keywords: cyber attack; attack prediction; attack projection; early-stage detection; Bayesian network

---

1. Introduction

Due to the constant development of cyber threats, various defense solutions need to be continuously improved. In addition to developing prevention systems, it is also necessary to focus on detection systems that help to obtain information about threats and attacks. The detection of malicious actions is one of the most critical cybersecurity issues. Intrusion detection refers to the detection of specific patterns or anomaly observations. Nowadays, however, we need to preventively anticipate upcoming harmful activities so that we can react to them and prevent an attack in time before it causes some damage.

Attack prediction study is not as prevalent as detection. Therefore, it is necessary to explore this area of interest because it is beneficial for the entire field of cybersecurity. To predict attacks, it is necessary to examine how they proceed and what steps are being taken. These data can be used to continually improve the systems to detect each phase of the attack. In this way, it is possible to detect the earlier stages of the attacks and predict how they proceed.

Early detection and prediction of cybersecurity incidents, such as attacks, is a challenging task. The threat landscape is continuously evolving, and even with the usage of intrusion detection systems, advanced attackers can spend more than 100 days in a system before being discovered [1]. After the detection of a security incident, we need to determine how the attack will proceed. This is essential because if we can stop the attacker in time, they cannot do as much damage.

It is important to learn from existing attacks so that we can develop tools to find out if such attacks have been repeated. Attack modeling is an intrusion-based methodology that allows one to focus on the different stages of an attack. It is aimed at focusing on different stages of attacks. By identifying attacks at different stages and by implementing tools to disarm the attacks at their various stages, one can take preventive measures to ensure that similar attacks will be detected. It is important to have a layered model to ensure that if one of the defense systems is bypassed, there is another defense line to protect one's organization's assets. That is why we need to establish a multi-layered model of cyber attacks.

Information **2020**, *11*, 560; doi:10.3390/info11120560　　　　　www.mdpi.com/journal/information

In recent years, it has not been sufficient to only be alerted of a security incident. Prevention of the attack altogether has become a necessity. The highest priority in computer security is to prevent an attack and stop the attacker from doing damage. If the path of an attack can be predicted, one has the ability to avoid attacks at every phase. By looking at a survey of the technology, from the host to the network level, one will have an opportunity to study tools or solutions that can be used in protecting against these threats. There are numerous existing prevention methods that are able to stop attacks in progress.

Recognizing an attack's steps is the goal of many cybersecurity analysts. The authors in [2] categorized prediction methods into three categories. An overview can be seen in Table 1.

Table 1. Overview of prediction categories.

Category	Description	Previous Surveys
Attack projection	What is an adversary going to do next?	Yang et al. [3]
Attack intention recognition	What is the ultimate goal of an adversary?	Ahmed and Zaman [4]
Attack/intrusion prediction	What type of attack will occur, when, and where?	Abdlhamed et al. [5]
Network security situation forecasting	How is the overall situation going to evolve?	Leau and Manickam [6]

The research is focused on early-stage detection and it is based on attack prediction, especially attack projection. This area focuses on the prognosis of the future steps of the attack. The projection of the future stages of an active cyber attack is essential in the context of Cyber Situational Awareness. The attacks often occur over an extended period of time. They involve a lot of steps and use multiple techniques for reconnaissance, exploitation, and obfuscation activities to achieve the attacker's goal. Therefore, it is not sufficient to just detect new or ongoing threats. The projection of future attack steps is deduced from already detected malicious activities. The estimates of current attack tactics may be used to assess imminent threats to critical assets [3].

This paper is based on the previous research of [7] and further develops research conducted by Ramaki et al. [8]. Based on the above-mentioned considerations, we state the following research sub-goals:

1. To propose a multi-stage model suitable for attack projection and early-stage detection, and
2. to design a model for early-stage detection of a cyber attack.

This paper is divided into seven sections. In Section 2, which is focused on related work and existing methods, the analysis of the current approaches of cyber attack prediction is provided. Section 3 presents the drawbacks of existing models and describes the suitable cyber attack model in detail. Subsequently, in Section 4, we propose the approach for early-stage detection of cyber attacks. This includes all of the necessary steps for data processing, alert aggregation, and causal relationship discovery. This section also covers the definition of Bayesian networks. After that, the model for the construction of the Bayesian network and prediction of cyber security alerts is proposed. Section 5 focuses on preprocessing and analysis of the data collection, including the creation of cyber alerts. The example cases of methods for aggregation, causal relationship discovery, and Bayesian network construction are shown. Section 6 presents and discusses the results of the presented methods. Concurrently, it describes groups of alerts and some of the attack paths. In the last section, the conclusion is provided.

2. Related Works

A large number of cyber attack prediction methods use discrete models and graph models, such as attack graphs, Bayesian networks, or Markov models.

In 1998, an attack graph was introduced by Swiler and Phillips [9]. It is a graphical representation of an attack scenario, and it has happened to be a popular method for formal description of attacks. It has become a foundation for other approaches, e.g., methods using Bayesian networks, Markov models, and game-theoretical methods. Their goal was to create a tool for qualitative and quantitative assessment of vulnerabilities. The approach was a great success because it examined a network security state from the system perspective.

Cao et al. [10,11] proposed another variant of the attack graph—the factor graph. It is a probabilistic model that consists of random variables and factor functions. In this paper, it is compared to Bayesian networks and Markov random fields. They used the factor graph to predict attacks with an accuracy of 75% over a dataset of actual security incidents (several years of reports).

The RTECA (Real-Time Episode Correlation Algorithm) was proposed in 2014 by Ramaki et al. [12]. It can be used to detect and predict multi-step attack scenarios. They explain the theoretical and functional implications of the creation of such a tool. Although they propose leveraging the attack graph, the authors have widely used causal correlations in their method.

The authors in [13] developed a method for correlating the intrusion alerts. It produces correlation graphs, which they use for creating attack strategy graphs. They presented techniques for automatically learning attack strategies from alerts raised by intrusion detection systems. These methods extracted attributes relevant to determining an attack strategy, which is represented as a directed acyclic graph, which they called an attack strategy graph. The nodes are known attacks, and the edges between them represent the order of attacks and relationships between them. They also developed a method for easier computer and network forensic analysis. It measures the similarity between sequences of alerts based on their strategies. Their research showed that the proposed methods can successfully extract invariant strategies from alert sequences and can also determine the likeness of those sequences. It can be widely used in identifying attacks that could have been missed by detection systems.

In [14], Li et al. presented another approach based on attack graphs. They described the generation of attack graphs constructed on a data mining approach. The algorithm they proposed uses association rule mining to get multi-step attack scenarios from Intrusion detection system (IDS) alert database. After that, the attack graph is created. The method is also used for calculating the predictability of the attack scenario. It is used for ranking the real-time detection and can help with intrusion prediction.

Liu and Peng [15] developed a game-theoretic framework used for attack prediction. The proposed method can quantitatively predict the probability of attack actions. It can also predict the strategic behavior of the attacker. Thus, it can optimize the precision of correlation-based prediction. This paper presents the first complex framework for motive-based modeling and inference of attackers' intents. In conclusion, the goal of this method is modeling and inference of attack intents, objectives, and strategies.

Wu et al. [16] used another attack prediction method using Bayesian networks. These methods are related to approaches based on attack graphs because a Bayesian network is built from an attack graph. The distinct characteristic of Bayesian networks is the conditional variables and probabilities that are considered in the model.

A Bayesian network is a probabilistic graphical model that describes the variables and the relationships between them. The network is a directed acyclic graph (DAG), where nodes represent the discrete or continuous random variables and edges depict the relationships between them. Each variable has a finite set of mutually exclusive states. The variable and direct edge form a DAG. To each variable A with parents $B_1, B_2, ..., B_n$, there is attached a conditional probability table $P(A|B_1, B_2, ..., B_n)$ [2].

Ishida et al. [17] proposed forecast techniques for fluctuation of attacks. They used Bayesian inference for calculating the probability of increase or decrease of the attacks. Two algorithms were considered in this paper—focusing on the attack cycle and the fluctuation range of the number of events. Because the event counts of some attacks change frequently, the proposed algorithms based on Bayesian inference were used for predicting the probability, since it can calculate event counts directly. Subsequently, they implemented the forecasting system and tested it on real IDS events.

A real-time alert correlation and prediction framework was introduced by Ramaki et al. [8]. The system includes an online and offline mode. In online mode, the attacker's next move is predicted by the Bayesian attack graph. In the offline mode, the Bayesian attack graph is constructed of low-level alerts. The authors used the DARPA 2000 dataset for research. The prediction accuracy was found to increase with the duration of the scenario for the attack. Thus, accuracy ranged from 92.3% when processing the first attack step to 99.2% when processing the fifth attack step.

Okutan et al. [18] used signals unrelated to the target network in their Bayesian-network-based attack prediction process. The signals include mention of Twitter attacks or the total number of Hackmageddon attacks [19]. As was shown in the results, the prediction accuracy differed from 63% to 99%, making it a promising method.

Since probabilistic graphical models are very powerful modeling and reasoning tools, Tabia et al. [20] proposed an efficient approach based on Bayesian networks. It allows the modeling of local influence relationships. It is dedicated to two main problems in alert correlation. Firstly, an approach based on Bayesian multi-nets was designed, which considered the local influence relationships to improve the prediction. The second problem occurs when multiple intrusion detection systems are in use in the network. In this case, too many of the raised alerts are redundant. Therefore, they proposed an approach for handling IDSs' reliability to reduce the number of false alerts. They based this approach on Pearl's virtual evidence [21].

Another widely used approach to predicting attacks is using Markov models. These methods were implemented along with approaches focused on attack graphs and Bayesian networks at the end of 2000. Farhadi et al. [22] proposed a complex system for alert correlation and prediction. Sequential pattern mining was used to collect the attack scenarios, which were then represented using the hidden Markov model, which was used to identify the attack strategy. Markov models perform well in the presence of unobservable states and transitions. They are not reliant on the possession of complete knowledge. This allowed a successful attack prediction, even though some of the attack stages were undetected or absent.

Using hidden Markov models, Sendi et al. [23] proposed a real-time intrusion prediction system. Multi-step attacks were the main interest in this paper. An empirical review showed how their method could anticipate multi-step attacks, which is especially useful in preventing the attacker from taking control of a huge number of hosts in the computer network.

In 2013, Shin et al. [24] introduced a probabilistic approach for the network-based intrusion detection system APAN, which uses a Markov chain for modeling unusual events in the network traffic to predict intrusion. Unlike other Markov-based methods, this method detects network anomalies and does not aim to predict the next step of an attack as different model-checking approaches do.

Holgado et al. [25] proposed a novel method based on a hidden Markov model for multi-step attack prediction using IDS alerts. They considered hidden states as a particular type of attack. At first, the preliminary training phase based on IDS alert information needs to be done. These observations are acquired by pairing the IDS alert information with a previously built database. Unsupervised and supervised methods for learning are performed in the training model. The prediction module can compute the best state sequence using the Viterbi and forward–background algorithms. The success of this method was shown in the successful detection of the distributed denial of service (DDoS) stages, which is a big problem in detection systems nowadays.

Table 2 shows the approaches in the cyber attack prediction methods. The first proposed method that has become popular involves prediction using an attack graph. It is the most transparent and easy-to-understand model for attack step representation. It has become beneficial in predicting the next steps in an attack. One of the lesser-known approaches is game theory. Nevertheless, it can be very useful in detecting DDoS attacks, which are very hard to predict. More commonly used methods include machine learning models. The first of them, the Bayesian network, has excellent accuracy results. However, it is tough to create this model from actual network traffic because the attackers can create loops in security alert data during attack implementation. Less intuitive approaches, but with great results, are the Markov chains and the hidden Markov model. These can be handy in predicting multi-step attacks.

Table 2. Summary of cyber attack prediction methods.

Paper	Approach/Model	Advantages and Limitations
Swiler and Phillips [9]	Attack graph	The first proposed methods
Cao et al. [10,11]	Attack graph	75% accuracy, factor graph
Ramaki et al. [12]	Attack graph	95% accuracy
Ning et al. [13]	Attack graph	Using correlation graphs
Li et al. [14]	Attack graph	Only partial graphs
Liu and Peng [15]	Game theory	Detection of nontrivial cyber attacks (e.g., DDoS)
Wu et al. [16]	Bayesian network	Only model extensions
Ishida et al. [17]	Bayesian network	70% accuracy, prediction of fluctuation of attacks
Ramaki et al. [8]	Bayesian attack graph	92.3–99.2% accuracy, real-time
Okutan et al. [18]	Bayesian network	63–99% accuracy, non-conventional signals
Tabia et al. [20]	Bayesian network	Reducing false alarms
Farhadi et al. [22]	Hidden Markov model	81.33–98.3% accuracy, data mining, illustrative example of a real-time attack projection
Sendi et al. [23]	Hidden Markov model	Prediction of the next step in a multi-step attack
Shin et al. [24]	Markov chain	Improving intrusion detection by predictions
Holgado et al. [25]	Hidden Markov model	Multi-step attack prediction, real-time, able to predict DDoS

On the other hand, Markov chains and the hidden Markov model need specific information. Due to the lack of information provided from the specific type of dataset, it is not possible to determine the values of the observation probability matrix. It is not certain what the probability of an attack is based on an observable alert. Therefore, we have decided to use a Bayesian network to create a method for cyber attack prediction.

3. The Proposed Cyber Attack Model

Cyber attack modeling is an important issue for securing any network and can help save money, time, and other resources. There exist several techniques that are used to model and analyze cyber attacks. The important part of understanding how every cyber attack works is to comprehend the steps that an attacker makes in order to reach their target. The goal of these approaches is to understand cyber attack characteristics to provide better security for a system. To defeat cyber attacks, it is also important to comprehend the attacker's objectives and their means. Understanding the characteristics of attacks is paramount in creating a good security strategy. Attack modeling is important in gaining a perspective on how can a cyber attack be stopped in a coordinated manner.

We considered using one of the following three models for analysis and use in our paper—the kill chain model [26], the model presented in [27], and the Diamond model [28]. The cyber kill chain model defines the path of a cyber attack. In this seven-layered model, each layer is critical for the evaluation of the attack. There are seven stages of the traditional kill chain model—reconnaissance, weaponization, delivery, exploitation, installation, command and control, and acting on the objective. This model is based on the assumption that attackers will seek to penetrate the computer system in a sequential and progressive way.

A sample anatomy of a cyber attack was also presented by Bou Harb et al. in their paper about cyber scanning [27]. The anatomy of the attack consists of the following steps—cyber scanning, enumeration, intrusion attempt, elevation of privilege, performing a malicious task, deploying malware/a backdoor, deleting forensic evidence, and exiting.

The Diamond model is one of the models for intrusion analysis. In this model, an attacker targets a victim on two main occasions, rather than using a sequence of continuous steps like the kill chain. This model consists of four elements—adversary, infrastructure, capability, and the victim [28].

Based on the analysis of the presented models, it was concluded that none of them met the requirements. The first two models contain stages that cannot be detected by IDSs. Since the Diamond model is not focused on attack steps, it is also not relevant to this research. That is why a new model needs to be developed. We introduce a hybrid model that includes four stages. This model can be seen in Figure 1.

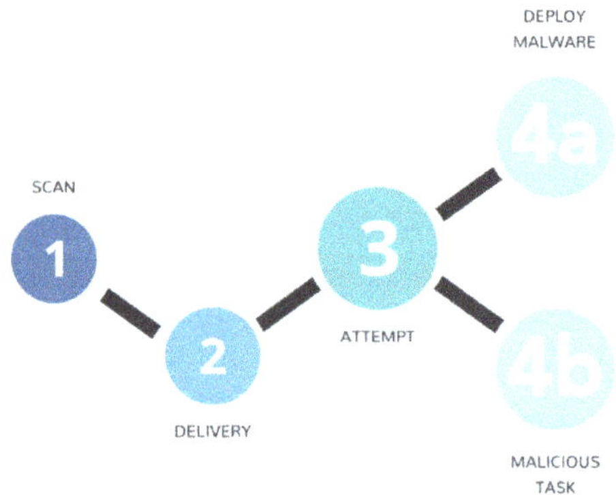

Figure 1. Proposed cyber attack model.

3.1. Scan

Cyber scanning is the first step in any sophisticated attack. This step is needed so the attacker can obtain information about their target, e.g., harvesting email addresses and login credentials or finding network vulnerabilities, etc. There are a variety of existing methods that an attacker can use to achieve this goal. There are two types of scanning techniques—passive and active.

An attempt to gain information about a target system or computer network that can be collected without actively engaging with the system is called passive scanning [29]. This can be performed by looking up the information about employees on a company's website. These can be email addresses, personal social media accounts, or phone numbers. LinkedIn and other social media networks can store the information of employees. This can help an attacker to identify their potential goal. In addition, social media accounts of employees can provide information about technologies used by a company. After finding out enough information about a victim, the possibility of success using social engineering techniques increases. Passive scanning is the most difficult thing to detect from the perspective of intrusion detection systems.

Active scanning is an attack in which an attacker engages a targeted network to gain information about vulnerabilities in it [29]. If an attacker is using an automated tool for network scanning, the IDS is likely to detect it and raise an alert. Performing active scanning is very valuable for determining any vulnerabilities that can be used. Network probing can be detected by correlating logs over a period of time. Therefore, it can be determined who may be targeting the system. This paper is focused on network scanning captured by Snort IDS. It can easily detect this type of stage. For example, if an attacker is using the NMAP tool to obtain information about a computer system (open ports or type of operating system), Snort can recognize a large number of various types of incoming packets and, therefore, identify the type of scan. It will raise a network-scan type of alert after recognizing this stage.

3.2. Delivery

Delivery is a critical part of every cyber attack model because it is responsible for an effective cyber attack. In most cyber attacks, it is necessary to have some kind of user cooperation, like downloading and executing malicious files or visiting malicious web pages on the internet. This stage presents a high risk for the attacker because delivery leaves evidence. Multiple delivery methods can be used, such as email attachments, phishing attacks, drive-by download, USB/removal media, or DNS cache poisoning [30].

Snort can detect malicious code by recognizing the transmission of executable code or suspicious strings in network traffic.

3.3. Attempt

Intrusion detection means discovering that some entity—an attacker—has attempted to gain—or has already gained—unauthorized access to the computer system. An intrusion attempt has a potential for a deliberate unauthorized attempt to enter a computer, system, or network to access information and manipulate information or render a system unreliable or unusable [31]. Intrusion attempts are experienced by victims, servers, networks, systems, and computers. These attempts can be discovered by intrusion detection systems. In the best case, it can be a false alarm because detection systems can sometimes raise false positive alerts. In order to determine if this was the case, it is needed to look at the details of the alert. If the attempt came from an infected system in the local network, it could provide information about this system; for example, the IP address that caused the alert. It can be later checked for any malicious activity. The last possibility is that there was an attempt to attack from an outside local network, but it was blocked. There is no way to determine if the attacker did not obtain any information. Detection of intrusion attempts can be helpful in defending a network, for example, by blacklisting IP addresses or updating firewall configurations.

3.4. Deploying Malware/Malicious Tasks

This stage contains the last four stages in the kill chain model [26]. In this phase, the malware is successfully installed on a computer system, or an attacker has obtained rights on the targeted device and is performing some malicious action. It starts with exploitation, which is initiated by installing the malware inside the target computer. The malware or the attacker has the required access rights. If the malware is an executable file or the malicious activity is based on code injection or an insider threat, then the installation is not required. After the malware is installed, it will start communication with the command and control server, which can be an attacker's device, server, or even social media network web server. If the attacker has gained access to a targeted computer system, he/she performs some malicious task; for example, stealing private and intellectual data from the network.

4. The Proposed Model for Early-Stage Detection of Cyber Attacks

This paper focuses on early-stage detection of cyber-attacks and, at the same time, the prediction of the subsequent stages of the attack. These attacks consist of multiple stages and may occur over an extended period of time. In this paper, we study how probabilistic inference can be used to analyze attack scenarios based on the information of the relations between alerts. This section describes a machine learning approach using a Bayesian network to predict cyber attacks' next steps. Algorithms for aggregation, causal relationship discovery, and Bayesian network construction will be introduced in this section.

4.1. Alert Aggregation

The first step in alert processing is aggregation, since intrusion detection systems are susceptible to alert flooding, meaning that they generate a huge number of alerts. Therefore, it is often hard to cope with a big amount of data. This issue can be solved by aggregating all of the alerts. Every aggregated event consists of:

- alert message,
- source IP address,
- destination IP address,
- source port,
- destination port,
- alert counter,
- start timestamp,
- end timestamp.

It is difficult to obtain a bigger picture in a large number of probes. Aggregation reduces the number of redundant alerts generated. This simplifies alert analysis and further processing. Possibly, this will not affect the information obtained in reduced alerts because only alerts that have the same important attributes are merged. Therefore, thousands of generated alerts are aggregated into a hyper-alert. Alerts were aggregated into one based on multiple attributes. All of the subsequent properties have to be met in two alerts for them to be merged into one:

- Time window—alerts have to be in the same time window, which was set to 10 min;
- Alert message and type—both alerts need to have the same message and, therefore, the same type;
- IP addresses—source and destination IP addresses have to be the same in both alerts.

The output of aggregation is stored in a multi-dictionary object: agr_alerts_all. The alerts are added to this data structure based on finding the key, which contains an alert message, source IP address, and destination IP address of the alert. The pseudo-code of the algorithm (Algorithm 1) can be seen below.

Algorithm 1 Aggregation of alerts.

Input *A* (all alerts generated by Snort)
Output *agr_alerts_all*

1: **procedure** AGGREGATION
2:     **for each** alert *a* in *A* **do**
3:         **if** *a* **not in** *agr_alerts_all* **then**
4:            *agr_alerts_all.add(a)*
5:         **else**
6:            **for each** alert *agr* **in** *agr_alerts_all* with same message, IP as *a* **do**
7:                **if** *a* and *agr* **in** same time window **then**
8:                    *agr.ports.add(a_{srcport}, a_{dstport})*
9:                    $agr.count \leftarrow agr.count + 1$
10:                **else**
11:                    *agr_alerts_all[agr].add(a)*
12:                **end if**
13:            **end for**
14:         **end if**
15:     **end for**
16: **end procedure**

4.2. Causal Relationship Discovery

In a causal Bayesian network, each arc is interpreted as a direct causal influence between a parent node and a child node relative to the other nodes in the network [32]. Therefore, after aggregation, the relationships between hyper-alerts are created. A directed edge between hyper-alerts A and B depicts that B occurred after A, and the event in A influences the event in B. This paper assumes that the attacker will not return to earlier stages and, thus, will not repeat certain steps. If the attacker was allowed to go back, it would not be possible to determine the probability deterministically. This would mean that there would be no final number of vertices where the attack could end. Therefore, it is expected that the attacker will end somewhere and, thus, will not perform the same stages over and over. Accordingly, a directed edge between A and B *(A -> B)* is constructed if the following criteria are met:

1. Time window—hyper-alerts A and B have to be within the same time window, which was set to 10 min;
2. Order—B occurred after A;
3. Stages—B is in a lower phase of attack than A;
4. IP addresses:

   - The source and destination IP addresses in A and B are the same, or
   - The destination IP address in A is the same as the source IP address in B.

After discovering all causal relationships between the hyper-alerts, the conditional probabilities need to be determined. The probability of *A -> B* or the probability of B under the condition of A will be denoted with the notation $P(B|A)$. However, before that, a correlation matrix showing the frequency of occurrences of the relationship *A -> B* needs to be created. The given matrix is stored in a pandas DataFrame object—correlated_alerts.

Algorithm 2 Causal relationship discovery.

Input *agr_alerts_all*
Output *correlated_alerts*

procedure CORRELATION
   for each alert *a* **in** *agr_alerts_all* **do**
      for each alert *b* **in** *agr_alerts_all* **do**
         if $a_{message}$!= $b_{message}$ and a_{phase} != b_{phase} **then**
            if *a* and *b* **in** same time window **then**
               if a_{srcIP}, a_{dstIP} == b_{srcIP}, b_{dstIP} **then**
                  if *a* before *b* and a_{phase} < b_{phase} **then**
                     correlated_alerts[*a*][*b*] ← *correlated_alerts*[*a*][*b*] + 1
                  else if *b* before *a* and b_{phase} < a_{phase} **then**
                     correlated_alerts[*b*][*a*] ← *correlated_alerts*[*b*][*a*] + 1
                  end if
               else if a_{dstIP} == b_{srcIP} and *a* before *b* and a_{phase} < b_{phase} **then**
                  correlated_alerts[*a*][*b*] ← *correlated_alerts*[*a*][*b*] + 1
               else if b_{dstIP} == a_{srcIP} and *b* before *a* and b_{phase} < a_{phase} **then**
                  correlated_alerts[*b*][*a*] ← *correlated_alerts*[*b*][*a*] + 1
               end if
            end if
         end if
      end for
   end for
end procedure

Subsequently, two variables representing causal relationships are defined: *Cor(A,B)*, which refers to the number of all causal relationships between alerts *A* -> *B*, hence the value in matrix: correlated_alerts[A][B], and *Cor(A,*)*, which represents the number of all causal relationships between A and all of the other hyper-alerts, which is the sum of the row A in the correlated_alerts matrix. Now, the conditional probabilities $P(B|A)$ can be computed as:

$$P(B|A) = \frac{Cor(A, B)}{Cor(A, *)} \; .$$

(1)

We work with table causal_relationships, which shows the results of the causal relationship discovery algorithm (Algorithm 2). It contains the following columns:

- Message of alert A,
- Index of alert A,
- Message of alert B,
- Index of alert B,
- Cor(A,B),
- Cor(A,*), and
- P(B | A).

All the data that we need for the construction of the network are in this table.

4.3. Bayesian Network

Bayesian networks form an important part of artificial intelligence by combining areas of probability theory and graph theory to solve uncertainty and complexity problems. They are one of the graphical probabilistic models, allowing a compact representation of the probability distribution of the simultaneous occurrence of monitored events [33]. The advantage of Bayesian networks is their relative intuitiveness for humans, since it is easier to understand the direct relationships between events and local probability distributions than the resulting probability distributions of occurrence of multiple events simultaneously. Their name is based on the Bayesian statistics on which they are based, which are based on the so-called Bayes' theorem expressing the change of previous assumptions in the light of new facts.

A Bayesian network (or causal network) is represented as a directed acyclic graph (DAG). Each node in this graph is a variable that has certain states. The directed edges in this graph represent relationships between variables—if there is a directed edge between two variables, then one is dependent on another [34]. If there is no connection between the two nodes, it does not mean that they are completely independent, because they can be connected through other nodes. However, they may become dependent or independent depending on the evidence that is established at other nodes. Nodes and connections build the structure of the Bayesian network, and we call it the structural specification.

This model consists of several parameters. The first one is the prior probability of parent nodes, which are not dependent on any states. Every child node has a conditional probability table (CPT) that states the prior knowledge between the node and its parent. An element of the CPT is defined by [35]:

$$CPT_{ij} = P(childstate = j | parentstate = i) \tag{2}$$

Some variables may have specific values that were observed. Let Y be the set of observed variables and Y_0 the corresponding set of values. Let X be the set of variables that are interesting for us. Inference is the process of computing the posterior probability $P(X|Y = Y_0)$. The posterior probability $P(X|Y = Y_0)$ is defined by [36]:

$$P(X|Y = Y_0) = \frac{P(X, Y = Y_0)}{P(Y = Y_0)} \tag{3}$$

In this paper, we will only consider a finite set $U = \{X_1, ..., X_n\}$ of discrete random variables where each variable X_i may take on values 0 or 1. In our case, these random variables will be aggregated alerts—hyper-alerts. We define each node of the causal network to have a binary state, i.e., 1 or 0. The value of 1 represents the alert in the node being raised, while the value of 0 indicates that it was not. Since the variables are discrete, the conditional probability tables contain the probabilities that a variable will contain one of all possible values for each combination of their parents' values.

4.4. Bayesian Network Construction

For Bayesian network generation, the causal relationships and conditional probabilities between hyper-alerts are needed. Therefore, the data from the previous subsection that are captured in the causal_relationships table will be used. The procedure for creating the Bayesian network is shown below. The method for creating the Bayesian network takes three input parameters: (I) agr_alerts_all, (II) correlated_alerts, and (III) causal_relationships. The output of this process is the Bayesian network of the hyper-alerts. The steps for Bayesian network construction are as follows:

1. Step 1: For each causal relationship between hyper-alerts x_i and x_j in the correlated_alerts table, a directed edge $x_i \rightarrow x_j$ is generated.

2. Step 2: Each node in the Bayesian network has a conditional probability table (CPT). This table displays the probability of the node given the values of its parent nodes. After all the edges are added to the network in Step 1, a method for generating the conditional probability table of each node was used. This method was introduced by [8]:

$$P(x_j|Pa(x_j)) = \begin{cases} 0, & \forall x_i \in Pa(x_j), x_i = false \\ P(\bigcup_{x_i=true} t_i) = 1 - \prod_{x_i=true}(1 - P(t_i)), & otherwise \end{cases} \quad (4)$$

In this function, x_i indicates values of the variables in the parent nodes $Pa(x_j)$ of x_j. The variable t_i denotes the transition that changes the state of the network from x_i to x_j, where $x_i \in Pa(x_j)$. Using this formula, each field in the conditional probability table in the vertex x_j is calculated based on the values in the parent nodes. It follows from this procedure that if all of the parent nodes have a value of 0 and, thus, no such alerts have occurred, the child node cannot occur either, and the value in the appropriate field will be 0. As a result of this procedure, the Bayesian network with conditional probability tables conditioned on different states of parents in each node has been constructed.

4.5. Alert Prediction

Based on the created Bayesian network, we can further calculate the probability of occurrence of a certain alert under the condition of other alerts using Bayesian inference. After an intrusion system generates one or more alerts, the so-called posterior probability, which is the probability of occurrence of every other state, can be computed. In this paper, only the probabilities of the alerts that belong to the third (Attempt) or fourth stage (Malicious) of the proposed model will be computed. Therefore, the most probable endings of the attack can be determined. With this knowledge, the system administrator can make precautions so they can mitigate the risk of a successful cyber attack that would otherwise cause bigger damage.

5. Data Preprocessing and Data Analysis

5.1. Datataset

For this paper, we work with the Intrusion Detection Evaluation Dataset (CICIDS2017) [37]. It includes benign and the most common attacks, which match the real-world data in (pcap format). It also includes the results of the network traffic analysis using the CICFlowMeter with labeled flows based on the timestamp, source, and destination IPs, source and destination ports, protocols, and attack. The data capturing period started at 9 a.m., Monday, 3 July 2017, and ended at 5 p.m. on Friday, 7 July 2017, for a total of five days. Monday only included regular traffic. The implemented attacks include Brute Force FTP, Brute Force SSH, DoS, Heartbleed, Web Attack, Infiltration, Botnet, and DDoS. They were executed in both the morning and afternoon on Tuesday, Wednesday, Thursday, and Friday [38]. We can see all of the implemented attacks for each day of data collection in Table 3.

Table 3. Overview of the activities in the dataset by day.

Day	Activity
Monday	Benign (Normal human activities)
Tuesday	Brute Force FTP (FTP-Patator)
	Brute Force SSH (SSH-patator)
	NAT Process on Firewall
Wednesday	DoS/DDoS
	DoS Slowloris
	DoS Slowhttptest
	DoS Hulk
	DoS GoldenEye
	NAT Process on Firewal
	Heartbleed Port 444
Thursday	Web Attack—Brute Force
	Web Attack—XSS
	Web Attack—SQL Injection
	NAT Process on Firewall
	Infiltration—Dropbox download
	Meta exploit Win Vista
	Infiltration—Cool disk—MAC
Friday	Botnet ARES
	Port Scan
	NAT Process on Firewall
	DDoS LOIT

This dataset was processed by the Snort intrusion detection system [39], which raised multiple alerts that matched the described attacks. As will be described later in this work, the alerts were aggregated based on the exact criteria into one hyper-alert. They were afterwards correlated—relationships between Hyper-alerts were established based on some properties. After these methods, a directed weighted graph was made, where vertices are hyper-alerts and the weighted edges are relations between them. This graph will be used as a base graph for the cyber attack prediction method using the discrete model—the Bayesian network.

The alerts generated from Thursday's traffic are used as a demonstration of a specific case. This day included multiple web attacks, such as cross-site scripting and SQL injection. Then, in the afternoon, an infiltration was performed. We can see the detailed description in Table 4.

5.2. Alert Preprocessing

For alert processing, the intrusion detection system Snort [40] was used. It analyzes the above-mentioned dataset based on two sets of alert rules. It takes a *.pcap* file as an input and resolves it based on the rules defined by the user in the configuration file. The first rule set was created by Hansson [41]. The company Proofpoint provides the second open rule set [42]. Rules from the first rule set are marked as "NF" and rules from the second rule set are marked as "ET". Once an attack or an abnormal situation is identified, an alert will be raised.

Snort [39] is an open-source network intrusion detection system that is capable of performing real-time traffic analysis and packet logging on IP networks. It can perform protocol analysis and content searching/matching, and can be used to detect a variety of attacks and probes, such as buffer overflows, stealth port scans, Common Gateway Interface (CGI) attacks, SMB probes, OS fingerprinting attempts, and much more.

Table 4. Description of Thursday's activities in the Intrusion Detection Evaluation Dataset (CICIDS2017).

Type of Attack	Time Interval	Notes
Web Attack —Brute Force	9:20 a.m.–10 a.m.	Attacker: Kali, 205.174.165.73, Victim: WebServer Ubuntu, 205.174.165.68
ine Web Attack—XSS	10:15 p.m.–10:35 a.m.	Attacker: Kali, 205.174.165.73, Victim: WebServer Ubuntu, 205.174.165.68
ine Web Attack—SQL Injection	10:40 a.m.–10:42 a.m.	Attacker: Kali, 205.174.165.73, Victim: WebServer Ubuntu, 205.174.165.68
ine Infiltration—Dropbox download	14:19 p.m–14:21 p.m., 14:33-14:35 p.m.	Attacker: Kali, 205.174.165.73 Victim: Windows Vista, 192.168.10.8
ine Infiltration—Cool disk—MAC	14:53 p.m.–15:00 p.m.	Attacker: Kali, 205.174.165.73 Victim: MAC, 192.168.10.25
ine Infiltration—Dropbox download	15:04 p.m.–15:45 p.m.	
(the first step)		Attacker: Kali, 205.174.165.73, Victim: Windows Vista, 192.168.10.8
(the second step)		Attacker: Vista, 192.168.10.8, Victim: All other clients

As mentioned above, the .pcap files from the CICIDS2017 dataset were analyzed by the Snort intrusion detection system. An output that contains raised alerts was generated into a .csv (comma-separated values) file. Data from this file were inserted into pandas DataFrame objects. Existing intrusion detection systems sometimes generate many false alerts. Therefore, we looked at the benign activity in the network from Monday. Data from each day that contained cyber attacks were filtered with benign activity from Monday's data. Thus, they would contain false positives. That being the case, only suspicious, non-normal cyber alerts were analyzed further. Based on the proposed model, which contains four stages, alerts from the intrusion detection system were assigned into various stages. Raised alerts were sorted based on their types to the proposed stages.

5.3. Alert Generation and Preprocessing

After using the Snort intrusion detection system to analyze these data, the input file in .csv format was created. Next, the data from this file were imported into Python, into the pandas DataFrame data structure. The data were filtered with alerts from Monday so they would not contain any false alerts or false positives. After that, 35 types of alerts remained. As we can see in Table 5, their numbers were very high.

5.4. Alert Aggregation

Therefore, aggregation must be performed on the data. This will significantly reduce the number of redundant alerts so that they can be further processed. As mentioned, an `AgrAlert` class object was created for each alert. The time window for aggregation was set to 10 min. The set of rules determined if the alerts would be merged into one. Two alerts had to be in the same time window and have the same message and the same pair of IP addresses. All of the assembled alerts were stored in the `agr_alerts_all` variable. After this procedure, only 203 hyper-alerts remained for further processing. A preview of them can be seen in Table 6.

5.5. Causal Relationship Discovery

A very important step in creating the Bayesian network is to determine the relationships between hyper-alerts. Therefore, an adjacency matrix `correlated_alerts` with frequencies on the edges was calculated based on a set of rules mentioned in the previous section. The value on the edge depicts how many times one alert occurred after another. A graph was later constructed from this matrix. Number one was added to the field in the matrix, and thus, a directed edge was created between the two alerts if the criteria between two alerts were fulfilled. They must have occurred in the same time window, which was set to an hour and a half. There were four rules that needed to be fulfilled. The first one was that the order had to be maintained. The second one said that the pair of IP addresses was the same or the destination IP address of the first alert was the same as the source IP address of the second one. The other two rules are based on the design of the attack model. The edge A -> B—hence, the edge from A to B—was only created if A and B were not in the same phase, and on top of that, the phase of A had to be in a lower phase than the alert B. In consequence, all of the cycles were removed, and a graph with a tree structure could be created. The table of the `correlated_alerts` can be seen in Table 7. In this table, we omitted the columns with only zero values (no correlation). The alerts in the columns are marked as follows: A3 - GPL NETBIOS SMB-DS IPCshare access; A12 - NF - Echo Reply Payload - Bigger than 100 bytes; A24 - NF - Bad TLD domain - click DNS query - Check domains; and A34 - NF - Echo Request Payload-Bigger than 100 bytes.

Table 5. Numbers of alerts after preprocessing.

Message	Count
NF—SCAN NMAP -sS 1024 Window	128,830
ET TROJAN Windows Microsoft Windows DOS prompt command Error not recognized	27,023
ET WEB_SERVER Script tag in URI Possible Cross Site Scripting Attempt	353
NF—SCAN nmap fingerprint attempt	346
NF—SCAN nmap XMAS	284
ET SCAN NMAP OS Detection Probe	254
NF—Echo Request Payload - Bigger than 100 bytes	254
NF—SCAN NMAP OS Detection Probe	254
NF—Bad TLD domain - click DNS query - Check domains	156
ET SCAN Suspicious inbound to MSSQL port 1433	110
ET SCAN Suspicious inbound to mySQL port 3306	108
ET SCAN Suspicious inbound to Oracle SQL port 1521	108
ET SCAN Suspicious inbound to PostgreSQL port 5432	107
NF—SCAN Nmap Scripting Engine User-Agent Detected (Nmap Scripting Engine)	96
ET SCAN Possible Nmap User-Agent Observed	96
ET SCAN Nmap Scripting Engine User-Agent Detected (Nmap Scripting Engine)	96
NF—Echo Reply Payload - Bigger than 100 bytes	88
GPL NETBIOS SMB-DS IPC$ share access	75
ET SCAN Potential VNC Scan 5800-5820	28
NF—SCAN Potential VNC Scan 5800-5820	28
ET SCAN Potential VNC Scan 5900-5920	28
ET WEB_SERVER Possible SQL Injection Attempt UNION SELECT	8
ET WEB_SERVER Possible SQL Injection Attempt SELECT FROM	6
ET POLICY Dropbox.com Offsite File Backup in Use	6
ET WEB_SERVER SELECT USER SQL Injection Attempt in URI	5
GPL ICMP_INFO PING *NIX	5
ET SCAN Behavioral Unusual Port 135 traffic Potential Scan or Infection	4
ET TROJAN Windows dir Microsoft Windows DOS prompt command exit OUTBOUND	3
ET WEB_SERVER Possible MySQL SQLi Attempt Information Schema Access	3
ET WEB_SERVER SQL Errors in HTTP 200 Response (error in your SQL syntax)	3
ET SCAN Behavioral Unusually fast Terminal Server Traffic Potential Scan or Infection (Inbound)	2
ET CURRENT_EVENTS Possible Phishing Redirect Dec 13 2016	2
ET POLICY SSLv3 outbound connection from client vulnerable to POODLE attack	2
ET SCAN Behavioral Unusually fast Terminal Server Traffic Potential Scan or Infection (Outbound)	2
NF—Possible Website defacement - Hacked by - Generic rule - Inbound	1

For the construction of Bayesian network, conditional probabilities between adjacent vertices needed to be calculated. Therefore, two variables were computed—`Cor(A, *)`, which is the number of all alerts originating in A, and `Cor(A, B)`, ergo, the number of times that alert B occurred after alert A. Next, the table of `causal_relationships` was created, containing all the information that will be used in Bayesian network creation.

Table 6. Aggregated alerts.

Message	IP Source	IP Destination
ET CURRENT_EVENTS Possible Phishing Redirect Dec 13 2016	104.23.129.81	192.168.10.17
ET CURRENT_EVENTS Possible Phishing Redirect Dec 13 2016	104.23.129.81	192.168.10.9
ET POLICY Dropbox.com Offsite File Backup in Use	162.125.4.5	192.168.10.8
ET POLICY Dropbox.com Offsite File Backup in Use	162.125.4.5	192.168.10.8
ET POLICY Dropbox.com Offsite File Backup in Use	162.125.18.133	192.168.10.9
...		
NF—SCAN nmap fingerprint attempt	192.168.10.8	192.168.10.12
NF—SCAN nmap fingerprint attempt	192.168.10.8	192.168.10.15
NF—SCAN nmap fingerprint attempt	192.168.10.8	192.168.10.50
NF—SCAN nmap fingerprint attempt	192.168.10.8	192.168.10.16
NF—SCAN nmap fingerprint attempt	192.168.10.8	192.168.10.14

Table 7. Table of correlated alerts.

Alert	A3	A12	A24	A34
A0—NF SCAN NMAP -sS 1024 Window	12	20	2	26
A1—ET SCAN Suspicious inbound to mySQL port 3306	10	16	2	24
A2—ET SCAN Possible Nmap User-Agent Observed	2	0	0	0
A6—ET SCAN Potential VNC Scan 5800-5820	6	12	2	18
A7—NF SCAN nmap XMAS	12	0	2	2
A8—ET SCAN Nmap Scripting Engine User-Agent Detected (Nmap)	2	0	0	0
A10—ET SCAN NMAP OS Detection Probe	12	0	2	2
A12—NF Echo Reply Payload - Bigger than 100 bytes	44	0	0	0
A13—NF SCAN nmap fingerprint attempt	12	0	0	2
A15—NF SCAN NMAP OS Detection Probe	12	0	2	2
A16—ET POLICY SSLv3 outbound connection vulnerable to POODLE attack	0	0	2	0
A19—ET SCAN Suspicious inbound to PostgreSQL port 5432	10	16	2	24
A21—ET SCAN Suspicious inbound to Oracle SQL port 1521	10	16	2	24
A26—ET SCAN Potential VNC Scan 5900-5920	6	12	2	18
A28—NF SCAN Potential VNC Scan 5800-5820	6	12	2	18
A31—ET SCAN Suspicious inbound to MSSQL port 1433	10	16	2	24
A33—NF SCAN Nmap Scripting Engine User-Agent Detected (Nmap)	2	0	0	0
A34—NF Echo Request Payload - Bigger than 100 bytes	12	0	2	0

5.6. Bayesian Network Construction

Since all of the information for making the Bayesian network was obtained, all that is left is to calculate conditional probability tables (CPTs) for each vertex in the graph. The created graph is illustrated in Figure 2. This graph will become the Bayesian network by adding the CPTs. The legend to the graph is shown in Table 8, joining numbers in the graph with alerts (equivalent to ID) and their messages and stages. Stages are numbered as follows: *1—SCAN, 2—DELIVERY, 3—ATTEMPT, 4a—DEPLOY MALWARE, 4b—MALICIOUS TASK.*

All of these alerts were assigned to these stages before. The keyword in the alert message is used to categorize a rule for intrusion detection. For example, most of the alerts that belong to the "Scan" phase contain the keyword "SCAN" in the beginning; therefore, they can be categorized mostly automatically. The alerts that could not be categorized automatically were assigned into stages manually.

Unfortunately, the table shows that no alerts belong to stages 3 (ATTEMPT) and 4a (DEPLOY MALWARE). The fourth phase is present only in 4b (MALICIOUS TASK) because there are no malware execution or distribution actions in this dataset. Due to its nature, there are no attack attempts on the Bayesian network. This is because this dataset was created by collecting and joining simple attacks. Therefore, there is no complex sample of an attack on it. Another reason is that Snort very often detected only an attempt phase, which was not related to other alerts. Therefore, a separate vertex without relationships with others was not added to the graph.

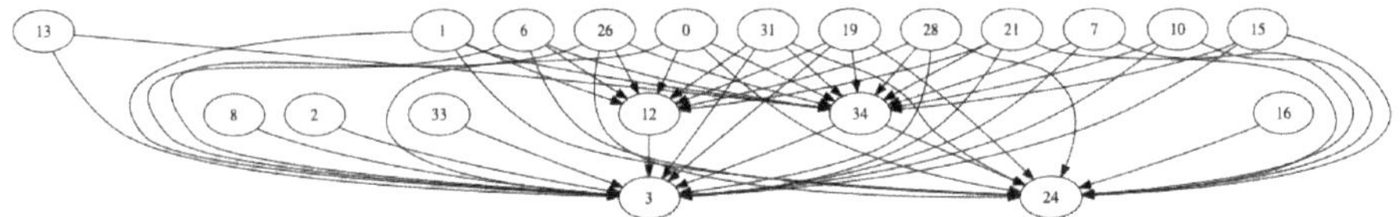

Figure 2. The graph that was created.

Table 8. Description of alerts.

ID	Message	Phase
A0	NF SCAN NMAP -sS 1024 Window	1
A1	ET SCAN Suspicious inbound to mySQL port 3306	1
A2	ET SCAN Possible Nmap User-Agent Observed	1
A3	GPL NETBIOS SMB-DS IPC$ share access	4b
A6	ET SCAN Potential VNC Scan 5800-5820	1
A7	NF SCAN nmap XMAS	1
A8	ET SCAN Nmap Scripting Engine User-Agent Detected (Nmap Scripting Engine)	1
A10	ET SCAN NMAP OS Detection Probe	1
A12	NF Echo Reply Payload - Bigger than 100 bytes	2
A13	NF SCAN nmap fingerprint attempt	1
A15	NF SCAN NMAP OS Detection Probe	1
A16	ET POLICY SSLv3 outbound connection from client vulnerable to POODLE attack	1
A19	ET SCAN Suspicious inbound to PostgreSQL port 5432	1
A21	ET SCAN Suspicious inbound to Oracle SQL port 1521	1
A24	NF Bad TLD domain - click DNS query - Check domains	4b
A26	ET SCAN Potential VNC Scan 5900-5920	1
A28	NF SCAN Potential VNC Scan 5800-5820	1
A31	ET SCAN Suspicious inbound to MSSQL port 1433	1
A33	NF SCAN Nmap Scripting Engine User-Agent Detected (Nmap Scripting Engine)	1
A34	NF - Echo Request Payload - Bigger than 100 bytes	2

The computational complexity of creating a Bayesian network is very high. Therefore, the algorithm will store the network's information, such as vertices and conditional probability tables, in the file. The output file is in BIF format, which is a structure that the `pgmpy` library can work with. We advanced through the vertices in topological order. Topological sorting is a linear ordering of its nodes such that for all directed paths from x to y $(x!=y)$, x comes before y in the ordering. Only the acyclic graph has a topological ordering.

6. Results and Discussion

In this section, the process of the evaluation of the individual methods will be presented. At first, the idea of this paper came from the research made by Ramaki et al. [8]. However, several problems occurred when we tried to follow the steps they took in their approach. The issues within that paper are presented in this section, as well as the modified approach that was developed in order to avoid those problems.

The authors in the mentioned paper tried to create a Bayesian network to predict cyber attack steps. After preprocessing our data from Snort, a similar method of aggregation was used to reduce the number of alerts. Next, the algorithm for causal relationship discovery was evaluated. It was also inspired by the method in the mentioned paper.

At this point, in their proposed method, a problem arose. The occurrence of cycles in the resulting graph was not mentioned in the given paper. However, we succeeded in solving this problem because the two conditions were stated in the causal relationship discovery phase. The attacker can only move across the stages of the cyber attack.

The creation of these conditions was based on the presented model of attack steps. The first condition resulted in eliminating the double-sided edges between the vertices. For example, many vertices from the SCAN phase would be interconnected. The rule said that there should be no edge between two hyper-alerts belonging to the same phase. The second rule resulted in the overall elimination of cycles. The rule was defined so that an oriented edge can only go from an alert that is in a lower phase to an alert in the higher phase. As a result, there are no cycles or bidirectional edges in the resulting graph and the Bayesian network.

After applying the rules, a Bayesian network was created. The following sections display an example of usage of the methods presented in the previous section. The data from Thursday were used to introduce the results of the individual implemented methods. This approach aims to create a Bayesian network designed to predict cyber attacks. The data from this day are described in detail. After that, the aggregation, causal relationship discovery, and Bayesian network construction methods are shown, along with other auxiliary methods.

After the Bayesian network was constructed and written into a file, the last step of the algorithm followed. The Python library pgmpy was used to load the network from the file with BIF format. After that, the Bayesian inference could finally be calculated, and the probabilities of attack steps could be computed. Our Bayesian network tells us how likely it is that an alert from the final stages will occur if the occurrence of an alert from the first phase, i.e., the scanning phase, has been detected. The results are presented in Table 9. In our case, there are two alerts in the Bayesian network belonging to the final phase. These two belong to phase number 4b, which symbolizes some malicious tasks.

The first of these alerts that we can see in the graph under ID - A3 is *GPL NETBIOS SMB-DS IPC$ share access*. This alert symbolizes the establishment of a connection using the samba protocol. It is meant to detect share access from outside the network. In this case, for example, it may be an EternalBlue infection. This exploit uses a vulnerability in Microsoft's implementation of the SMB (Server Message Block) protocol for remote code execution. However, this type of network activity is often a false positive case. It can be a null session attack on samba functionality, which enables anonymous access to hidden administrative shares on a system. On the other hand, it may be legitimate network traffic; for example, traffic to a domain controller. This alert emerges if the rule defined in the Snort detection system is fulfilled.

Table 9. The probability of reaching the final stage provided that specific stages occur.

Late Stage	Condition	Probability	Late Stage	Condition	Probability
A3	A13	0.9998	A24	A13	0.6989
	A1	0.9995		A1	0.7059
	A6	0.9995		A6	0.7081
	A26	0.9995		A26	0.7081
	A0	0.9996		A0	0.7049
	A31	0.9995		A31	0.7059
	A19	0.9996		A19	0.7059
	A28	0.9995		A28	0.7081
	A21	0.9995		A21	0.7059
	A7	0.9997		A7	0.7189
	A10	0.9997		A10	0.7189
	A15	0.9997		A15	0.7189
	A8	1.0000		A8	0.6985
	A2	1.0000		A2	0.6985
	A33	1.0000		A33	0.6985
	A16	0.9992		A16	1.0000

The second alert that belongs to the final phase is *NF - Bad TLD domain - click DNS query - Check domains*. It can be recognized under ID - A24 in the graph. This means that the device has accessed a domain whose TLD is marked as malicious. A top-level domain can be lablled as bad when there were some indications that it was tied to spam or malware dissemination. Websites using the new top-level domains, such as .men, .work, or .click, are some of the riskiest. The rule that was made for this detection is presented next.

The mentioned alerts and all others emerged from the rules defined in Snort IDS. As was mentioned before, two sets of rules were used—NF and ET.

Next, the types of paths in the graph will be described based on a certain grouping of alerts. Based on the groups of alerts from the first phase according to the type of attack, paths were created in the graph. Their probabilities were calculated using the Bayesian network. Three groups of alerts belonging to the SCAN phase were created.

6.1. SCAN Suspicious Inbound to SQL

The first group of alerts contains all of the SQL-related alerts. All of them cohere with SQL port scans. The alerts belonging to this group are:

- A1—msg: "ET SCAN Suspicious inbound to mySQL port 3306",
- A19—msg: "ET SCAN Suspicious inbound to PostgreSQL port 5432",
- A21—msg: "ET SCAN Suspicious inbound to Oracle SQL port 1521",
- A31—msg: "ET SCAN Suspicious inbound to MSSQL port 1433".

All of the alerts can proceed into both of the second-stage alerts. Alerts in this phase show that a payload of at least 100 bytes has been transmitted into and out of the network:

- A12—msg: "NF - Echo Reply Payload - Bigger than 100 bytes", and
- A34—msg: "NF - Echo Request Payload - Bigger than 100 bytes".

From this point, the attack can end in either one of the final stages. If the attack's path contained an alert with number 12, it will certainly end in an alert with number 3. If the path went through alert 34, the attack can end in any of the final stages:

- A3—msg: "GPL NETBIOS SMB-DS IPC$ share access", and
- A24—msg: "NF - Bad TLD domain - click DNS query - Check domains".

Table 10 shows the probabilities of occurrence of final stages based on some of the paths in the graph that contain the first and the second stages.

Table 10. Probabilities of paths including alerts from the group SCAN Suspicious inbound to SQL.

Observed Alerts		Final Stage Alert	Probability
The First Stage	The Second Stage		
A1	A12	A3	1.0000
A1	A34	A3	0.9998
A1	A12	A24	0.7105

After summarizing the results of these calculations, we get clear paths and ends of probable attacks. In the first phase, the SQL service (ports 3306,5432,1521,1433) is scanned from the external network. Subsequently, the payload (>100 bytes) is downloaded or sent. Next, the malicious activity has been executed—access from the external network to the samba protocol or demand for a bad TLD domain (e.g., .tk, .fit, or .rest [43]).

6.2. SCAN 1

The second group contains alerts that did not connect to all of the subsequent stages. All of the alerts only continued into one of the second-stage alerts. The contents of this group include:

- A7—msg: "NF—SCAN nmap XMAS",
- A10—msg: "ET—SCAN NMAP OS Detection Probe", and
- A15—msg: "NF—SCAN NMAP OS Detection Probe".

As mentioned, all of the alerts can only go to a second-stage alert with ID A34. This alert was described in the previous group. The message of this variable is 'NF - Echo Request Payload - Bigger than 100 bytes'.

From this point, the attack can proceed and end in any of the final stages. Both of them have also been described in the previous group's definition. An example of the attack path that contains the first and the second stage and ends in the fourth stage can be seen in Table 11.

Table 11. Probabilities of paths including alerts from the group SCAN 1.

Observed Alerts		Final Stage Alert	Probability
The First Stage	The Second Stage		
A7	A34	A3	0.9999
A7	A12	A24	0.7249

From these findings, it is possible to deduce the path by which an attacker can lead to their malicious activity. The first phase is scanning of some device. It can either be an NMAP scan that sets the FIN, PSH, and URG flags in the packet, or an NMAP scan that tries to detect the target operating system. After this, packets with payloads bigger than 100 bytes are downloaded or sent. The attack can end in two ways—connection to the samba protocol from an external network or requesting one of the bad TLD domains.

6.3. SCAN 2

This group of alerts are connected to every one of the subsequent stages. It contains NMAP scans or VNC scans. This group includes:

- A0—msg: "NF - SCAN NMAP -sS 1024 Window",
- A6—msg: "ET SCAN Potential VNC Scan 5800–5820",
- A13—msg: "NF - SCAN nmap fingerprint attempt",
- A26—msg: "ET SCAN Potential VNC Scan 5900–5920", and
- A28—msg: "NF - SCAN Potential VNC Scan 5800–5820".

As was described before, the attack from this group of alerts can proceed in more than one path. The second stage is the same as that in the SCAN Suspicious inbound to the SQL group. The attack can end in both of the mentioned final stages that were described before. An example of an attack that starts with scanning of the designated ports can be seen in Table 12.

Table 12. Probabilities of paths including alerts from the group SCAN 2.

Observed Alerts		Final Stage Alert	Probability
The First Stage	The Second Stage		
A0	A12	A3	1.0000
A0	A34	A3	0.9998
A0	A12	A24	0.7098

From these data, it was concluded that the attack could proceed as follows. The goal of the first phase of the attack was to scan a wide range of ports to find some open vulnerability. It could be scanning all of the designated ports (0–1024) so the attacker can find out what services run on the device and if any of them are vulnerable. The attacker can also be scanning Virtual Network Computing (VNC) ports, which belong to a graphical desktop sharing system. The second phase of the attack is once again sending or downloading big payloads of data, which is suspicious. The attack could end in any of the final stages. The more probable result is that the attacker found an open, vulnerable Server Message Block (SMB) port and tried to exploit this vulnerability.

Since the dataset was not complex and did not contain a wide range of data, further studies are needed in order to test this approach. A larger dataset with an emphasis on attack stages is needed so the proposed methods can be further verified. In this evaluation, we presented some of the attack paths that are likely to happen in real traffic. From the data that were collected, it was discovered that many of the attackers use malicious domains to execute an attack. It was also concluded that some of the ports, like SMB ports, that are considered vulnerable are very often the target of an attack.

7. Conclusions

Prompt response to security incidents means minimizing the damage caused by the security incidents to the organization. The main goal of the organization is maximum preparation for handling security incidents, as well as their prevention through proactive activities. The transition from reactive activities to proactive activities is currently a challenge in cybersecurity research.

The attack—or the steps of the attackers—can be divided into several stages. Consequently, one of the ways to move from reactive activities to proactive activities is to identify the initial stages of the attack. To be able to make such an identification, it is necessary to predict the next steps of the attacker, which is called the projection of attacks in the current research.

Within this paper, we focused on the projection of attacks, and for this purpose, we defined four stages of attacks. The aim of the paper was the prediction of the final stages (the third and fourth stages), provided that we knew the first two stages of the attack. For this purpose, we chose Bayesian networks.

The aim was to design a model that, based on the projection of the attack, would identify the early stages of the attack. The proposed model includes not only the aggregation of alerts, but also their correlation. We used the proposed attack model for the prediction itself. As research has shown, the paper [8] on which our research is based and was the inspiration for the use of the Bayesian network does not take into account the situation where an attacker can return within a particular stage. The situation creates a cyclic graph, which creates a problem, since the Bayesian network assumes an acyclic graph as the input. For this reason, it was necessary to add a condition to the model under which the attacker does not return within the individual stages.

We tested the proposed model on the publicly available Intrusion Detection Evaluation Dataset CICIDS2017. In the paper, we showed the real application of the proposed model on a prepared dataset, including the creation of alerts, their aggregation and correlation, and the subsequent detection of early stages based on the attack projection.

This research can be expanded in the future. There are several challenges for future work. One of them is the processing and prediction of events, even if there are cycles in the attack graph. This case is problematic because many computational models that take into account acyclic graphs cannot be used. Therefore, it could be appropriate to try another method. For example, using attack graph modeling or a hidden Markov model would be successful. The second challenge that this research presents is creating a complex and comprehensive dataset. Nowadays, to the best of our knowledge, no suitable datasets have been created that emphasize the attacks. Therefore, it is important to work with a dataset that would contain attacks that cover all detectable stages of an attack.

Author Contributions: Conceptualization, M.P., T.B. and P.S.; methodology, M.P. and T.B.; data processing, M.P. and T.B.; results analysis, M.T. and P.S.; writing—original draft preparation, M.P.; writing—review and editing, M.P., P.S. and T.B.; supervision, P.S. All authors have read and agreed to the published version of the manuscript.

Funding: This paper is funded by the project VVGS-PF-2019-1062, project VVGS-PF-2020-1427, and the Slovak APVV project under contract No. APVV-17-0561.

Conflicts of Interest: The authors declare no conflict of interest.

References

1. FireEye. Common Vulnerability Scoring System. 2018. Available online: https://www.fireeye.com/content/dam/collateral/en/mtrends-2018.pdf (accessed on 19 November 2020).
2. Husák, M.; Komárková, J.; Bou-Harb, E.; Čeleda, P. Survey of attack projection, prediction, and forecasting in cyber security. *IEEE Commun. Surv. Tutor.* **2018**, *21*, 640–660. [CrossRef]
3. Yang, S.J.; Du, H.; Holsopple, J.; Sudit, M. Attack projection. In *Cyber Defense and Situational Awareness*; Springer: Berlin/Heidelberg, Germany, 2014; pp. 239–261.
4. Ahmed, A.A.; Zaman, N.A.K. Attack Intention Recognition: A Review. *IJ Netw. Secur.* **2017**, *19*, 244–250.
5. Abdlhamed, M.; Kifayat, K.; Shi, Q.; Hurst, W. Intrusion prediction systems. In *Information Fusion for Cyber-Security Analytics*; Springer: Berlin/Heidelberg, Germany, 2017; pp. 155–174.
6. Leau, Y.B.; Manickam, S. Network security situation prediction: A review and discussion. In *International Conference on Soft Computing, Intelligence Systems, and Information Technology*; Springer: Berlin/Heidelberg, Germany, 2015; pp. 424–435.
7. Mézešová, T.; Sokol, P.; Bajtoš, T. Evaluation of Attacker Skill Level for Multi-stage Attacks. In Proceedings of the 2019 11th International Conference on Electronics, Computers and Artificial Intelligence (ECAI), Pitesti, Romania, 27–29 June 2019; pp. 1–6.

8. Ramaki, A.A.; Khosravi-Farmad, M.; Bafghi, A.G. Real time alert correlation and prediction using Bayesian networks. In Proceedings of the 2015 12th International Iranian Society of Cryptology Conference on Information Security and Cryptology (ISCISC), Rasht, Iran, 2–3 September 2015; pp. 98–103.

9. Swiler, L.P.; Phillips, C. *A Graph-Based System for Network-Vulnerability Analysis*; Technical Report; Sandia National Labs.: Albuquerque, NM, USA, 1998.

10. Cao, P.; Chung, K.W.; Kalbarczyk, Z.; Iyer, R.; Slagell, A.J. Preemptive intrusion detection. In Proceedings of the 2014 Symposium and Bootcamp on the Science of Security, Raleigh, NC, USA, 8–9 April 2014; p. 21.

11. Cao, P.; Badger, E.; Kalbarczyk, Z.; Iyer, R.; Slagell, A. Preemptive intrusion detection: Theoretical framework and real-world measurements. In Proceedings of the 2015 Symposium and Bootcamp on the Science of Security, Urbana-Champaign, IL, USA, 21–22 April 2015; p. 5.

12. Ramaki, A.A.; Amini, M.; Atani, R.E. RTECA: Real time episode correlation algorithm for multi-step attack scenarios detection. *Comput. Secur.* **2015**, *49*, 206–219. [CrossRef]

13. Ning, P.; Xu, D. Learning attack strategies from intrusion alerts. In Proceedings of the 10th ACM Conference on Computer and Communications Security, Washington, DC, USA, 27–30 October 2003, pp. 200–209.

14. Li, Z.; Lei, J.; Wang, L.; Li, D. A Data Mining Approach to Generating Network Attack Graph for Intrusion Prediction. In Proceedings of the Fourth International Conference on Fuzzy Systems and Knowledge Discovery (FSKD 2007), Haikou, China, 24–27 August 2007; Volume 4, pp. 307–311.

15. Liu, P. *A Game Theoretic Approach to Cyber Attack Prediction*; Technical Report; Pennsylvania State University: University Park, PA, USA, 2005.

16. Wu, J.; Yin, L.; Guo, Y. Cyber attacks prediction model based on Bayesian network. In Proceedings of the 2012 IEEE 18th International Conference on Parallel and Distributed Systems, Singapore, 17–19 December 2012; pp. 730–731.

17. Ishida, C.; Arakawa, Y.; Sasase, I.; Takemori, K. Forecast techniques for predicting increase or decrease of attacks using bayesian inference. PACRIM. In Proceedings of the 2005 IEEE Pacific Rim Conference on Communications, Computers and signal Processing, Victoria, BC, Canada, 24–26 August 2005; pp. 450–453.

18. Okutan, A.; Yang, S.J.; McConky, K. Predicting cyber attacks with bayesian networks using unconventional signals. In Proceedings of the 12th Annual Conference on Cyber and Information Security Research, Kowloon, Hong Kong, China, 15–16 August 2017; p. 13.

19. Passeri, P. HACKMAGEDDON, Information Security Timelines and Statistics. 2016. Available online: https://www.hackmageddon.com/ (accessed on 19 November 2020).

20. Tabia, K.; Leray, P. Bayesian network-based approaches for severe attack prediction and handling IDSs' reliability. In *International Conference on Information Processing and Management of Uncertainty in Knowledge-Based Systems*; Springer: Berlin/Heidelberg, Germany, 2010, pp. 632–642.

21. Pearl, J. *Probabilistic Reasoning in Intelligent Systems: Networks of Plausible Inference*; Elsevier: Amsterdam, The Netherlands, 2014.

22. Farhadi, H.; AmirHaeri, M.; Khansari, M. Alert correlation and prediction using data mining and HMM. *ISeCure* **2011**, *3*, 77–101.

23. Sendi, A.S.; Dagenais, M.; Jabbarifar, M.; Couture, M. Real time intrusion prediction based on optimized alerts with hidden Markov model. *J. Networks* **2012**, *7*, 311.

24. Shin, S.; Lee, S.; Kim, H.; Kim, S. Advanced probabilistic approach for network intrusion forecasting and detection. *Expert Syst. Appl.* **2013**, *40*, 315–322. [CrossRef]

25. Holgado, P.; Villagrá, V.A.; Vazquez, L. Real-time multistep attack prediction based on hidden markov models. *IEEE Trans. Dependable Secur. Comput.* **2017**, *17*, 134–147. [CrossRef]

26. Hutchins, E.M.; Cloppert, M.J.; Amin, R.M. Intelligence-driven computer network defense informed by analysis of adversary campaigns and intrusion kill chains. *Lead. Issues Inf. Warf. Secur. Res.* **2011**, *1*, 80.

27. Bou-Harb, E.; Debbabi, M.; Assi, C. A systematic approach for detecting and clustering distributed cyber scanning. *Comput. Networks* **2013**, *57*, 3826–3839. [CrossRef]

28. Caltagirone, S.; Pendergast, A.; Betz, C. *The Diamond Model of Intrusion Analysis*; Technical Report; Center For Cyber Intelligence Analysis and Threat Research: Hanover, MD, USA, 2013.

29. Bou-Harb, E.; Debbabi, M.; Assi, C. Cyber scanning: a comprehensive survey. *IEEE Commun. Surv. Tutor.* **2013**, *16*, 1496–1519. [CrossRef]

30. Yadav, T.; Rao, A.M. Technical aspects of cyber kill chain. In *International Symposium on Security in Computing and Communication*; Springer: Berlin/Heidelberg, Germany, 2015; pp. 438–452.

31. Available online: https://www.mycert.org.my (accessed on 19 November 2020).

32. Cooper, G. An overview of the representation and discovery of causal relationships using Bayesian networks. In *Computation, Causation, and Discovery*; MIT Press: Menlo, CA, USA, 1999; pp. 4–62.

33. Guo, H.; Hsu, W. A Survey of Algorithms for Real-Time Bayesian Network Inference. 2002. Available online: https://www.aaai.org/Papers/Workshops/2002/WS-02-15/WS02-15-001.pdf (accessed on 19 November 2020).

34. Friedman, N.; Geiger, D.; Goldszmidt, M. Bayesian network classifiers. *Mach. Learn.* **1997**, *29*, 131–163. [CrossRef]

35. Qin, X.; Lee, W. Attack plan recognition and prediction using causal networks. In Proceedings of the 20th Annual Computer Security Applications Conference, Tucson, AZ, USA, 6–10 December 2004; pp. 370–379.

36. Zhang, N.L.; Poole, D. A simple approach to Bayesian network computations. In Proceedings of the Biennial Conference-Canadian Society for Computational Studies of Intelligence, Banff Park Lodge, Banff, Alberta, 16–20 May 1994; pp. 171–178.

37. Sharafaldin, I.; Lashkari, A.H.; Ghorbani, A.A. Toward Generating a New Intrusion Detection Dataset and Intrusion Traffic Characterization. 2018. Available online: https://www.scitepress.org/Papers/2018/66398/66398.pdf (accessed on 19 November 2020).

38. Intrusion Detection Evaluation Dataset (CICIDS2017), U.o.N. Brunswick. 2017. Available online: https://www.unb.ca/cic/datasets/ids-2017.html (accessed on 19 November 2020).

39. Snort - Network Intrusion Detection & Prevention System. Available online: https://www.snort.org (accessed on 19 November 2020).

40. SNORT Users Manual 2.9.13. The Snort Project 2019. 2019. Available online: https://www.snort.org/documents/snort-users-manual (accessed on 19 November 2020).

41. Hansson, L. NF IDS Rules. Available online: https://networkforensic.dk/SNORT/ (accessed on 19 November 2020).

42. Emerging Threats rules. Available online: https://rules.emergingthreats.net/ (accessed on 19 November 2020).

43. What Is Cyber Attack? 2018. Available online: https://www.upguard.com/blog/cyber-attack (accessed on 19 November 2020).

 *information*

Review

Attacker Behaviour Forecasting Using Methods of Intelligent Data Analysis: A Comparative Review and Prospects

Elena Doynikova [1,*], **Evgenia Novikova** [1,2] **and Igor Kotenko** [1]

[1] St. Petersburg Institute for Informatics and Automation of the Russian Academy of Sciences,
   St. Petersburg 199178, Russia; novikova@comsec.spb.ru (E.N.); ivkote@comsec.spb.ru (I.K.)
[2] Saint Petersburg Electrotechnical University "LETI", Department of computer science and technology,
   St. Petersburg 197022, Russia
* Correspondence: doynikova@comsec.spb.ru

Received: 18 February 2020; Accepted: 16 March 2020; Published: 23 March 2020

Abstract: Early detection of the security incidents and correct forecasting of the attack development is the basis for the efficient and timely response to cyber threats. The development of the attack depends on future steps available to the attackers, their goals, and their motivation—that is, the attacker "profile" that defines the malefactor behaviour in the system. Usually, the "attacker profile" is a set of attacker's attributes—both inner such as motives and skills, and external such as existing financial support and tools used. The definition of the attacker's profile allows determining the type of the malefactor and the complexity of the countermeasures, and may significantly simplify the attacker attribution process when investigating security incidents. The goal of the paper is to analyze existing techniques of the attacker's behaviour, the attacker' profile specifications, and their application for the forecasting of the attack future steps. The implemented analysis allowed outlining the main advantages and limitations of the approaches to attack forecasting and attacker's profile constructing, existing challenges, and prospects in the area. The approach for attack forecasting implementation is suggested that specifies further research steps and is the basis for the development of an attacker behaviour forecasting technique.

Keywords: cyber attack; attacker; attacker profile; attacker behaviour; metrics; features; attributes; intelligent data analysis; attack forecasting; comparative review

1. Introduction

The attacker model plays an important role in the tasks of the attack modelling, forecasting, and risk analysis. Existing approaches consider different attacker's characteristics when modelling attacks. Some of them use high level goals of the malefactor [1]—hackers, spies, terrorists, corporate raiders, professional criminals, vandals, and voyeurs.

Others approaches analyze the location of the attacker—internal or external [2]—and the complexity of the vulnerabilities they exploit [3]—script kiddies, hackers, and botnet owners.

In [2], the classification of attackers based on several attributes is suggested. The analyzed parameters include the quantity of the malefactors, their motives, and their goals, which allows authors to define three types of attackers—individuals, organized groups, and intelligence agency.

Federal Service for Technical and Expert Control (FSTEC) of Russian Federation classifies attacker according to its skills and location in the system—internal attacker with low skills, internal attacker with medium skills, internal attacker with high skills, external attacker with low skills, external attacker with medium skills, and external attacker with high skills.

Various approaches are used to further clarify the type of attacker carrying out an ongoing attack. These approaches include the following:

1. Techniques based on attack graph analysis.
2. Techniques based on hidden Markov model.
3. Techniques based on fuzzy inference.
4. Techniques based on attributing cyber attacks using intelligent data mining techniques including neural networks, statistics, and so on.

However, it is possible to highlight several limitations of these techniques. One of them is the lack of a unified and validated approach to the attacker model description. According to these approaches, different attackers' attributes result in different attackers' profiles, and these approaches as a rule do not consider the latest paradigm shifts and novel attack vectors that appear owing to the development of the Internet of Things (IoT), cyber-physical systems, software defined networking (SDN), 5G mobile networks, and so on. Another significant problem in the attacker profiling process is the lack of consistent labeled datasets for model training.

There are currently a number of surveys in the area of attack forecasting and prediction. For example, in 2016, Gheyas and Abdallah [4] surveyed the detection and prediction of insider threats. In [5], the authors investigated the attack projection, prediction, and forecasting methods in cyber security. They distinguish between attack projection that relates to the next adversary steps [6]; attack intention recognition, which deals with detection of the final malefactor goal [7]; attack/intrusion prediction, which relates to the definition of which type of attack will take place, as well as when and where it will arise [8]; and, finally, network situation forecasting, which is connected with assessment of possible cyber security risks and their evolution. The authors outlined four different classes of the approaches based on the type of mathematical model used—discrete models (attack graphs, Bayesian networks, Markov Models), continuous models (time series analysis, Grey models), machine learning techniques and data mining techniques, and other approaches (similarity based, among others). They also focused on the problem of the source data used for predictions as different approaches operate on different levels of abstraction and require different types of data. They showed that the following types of input data can be used: (1) raw data, such as network traffic and system logs; and (2) abstract data, such as alerts from intrusion detection/protection systems and/or numerical representation of network security state. The authors discussed the advantages and limitations of each approach and showed the current status of each approach, that is, proof of concept or live tool. However, these surveys did not address the issues of the attacker profile definition or attacker attribution and its influence on the attack forecasting process.

Another interesting review of attacker models and profiles for cyber-physical systems (CPSs) is provided in [9]. The authors focused on the related work on the following: (1) attacks against CPS and ad-hoc attacker models, (2) profiling attackers for CPS, and (3) generic attacker models for CPS. They reviewed works that discuss attackers who target or leverage the physical layer in their attacks (mechanical, electrical interactions). The authors gave the main definitions concerning the attacker and attacker's profile. For example, they define an attacker as a person(s) aimed to achieve some malicious goal in the system, and an attacker profile as a template listing possible actions, motivations, or capabilities of the attacker. They note that an attacker model (together with compatible system models) should represent all possible interactions between the attacker and the system. Besides, they also include the constraints for the attacker model such as finite computational resources and no access to shared keys.

The authors reviewed 19 related works and came to the following conclusions:

1. Seven works explicitly use different attacker profiles, seventeen define dimensions, and the vast majority use actions to characterize the attacker. Just two works define a system model and perform risk analysis without explicitly considering an attacker model. This shows the trend of defining an attacker model to perform security analysis of CPS and, at the same time, there exist various ways to model the attacker.

2. All the papers share the same actions or the same intuitions on the attackers, but they apply those actions to different definitions of attacker models.

3. Different works propose different attacker profiles. The boundaries between the different attacker profiles are not well defined, thus it is hard to classify a specific attacker as one specific profile. The authors outline the following six types of attackers based on related research: (1) a basic user [10,11] (also known as script kiddie, unstructured hacker, hobbyist, or cracker) uses already established and potentially automated techniques to attack a system, and has average access to hardware, software, and Internet connectivity; (2) an insider [11–14] (disgruntled employees or social engineering victims) can cause damage to the target depending on the employment position or the system privileges he/she owns (e.g., user, supervisor, administrator)—this type is of high importance for systems that are mainly protected through air-gaps between the system network and the outside world (often used in CPS); (3) a hacktivist [10–12] aims to promote a political agenda, often related to freedom of information (e.g., Anonymous); (4) a terrorist [11–13], also known as cyber-terrorist, is a politically motivated attacker who uses information technology to cause severe disruption or widespread fear [15,16]; (5) a cybercriminal [10–14] (sometimes called black hat hacker or structured hacker) is an attacker with extensive security knowledge and skills, he/she takes advantage of known vulnerabilities, and potentially has the knowledge and intention of finding new zero-day vulnerabilities, his/her goals can range from blackmailing to espionage (industrial, foreign) or sabotage; (6) a nation-state [10–13] is an attacker sponsored by a nation/state, and his/her targets are usually public infrastructure systems, mass transit, power or water systems, and general intelligence.

4. Finally, the authors outlined nine common parameters that are used to generate metrics. Examples of metrics are as follows:

    a. tools (resources) available, also known as attacklets, or actions in the abstract definition of the attacker model—these define which types of tools are available to the attacker;

    b. camouflage or preference to stay hidden—expresses the aim and/or the ability of the attacker to not be tracked down after or while performing an attack;

    c. distance to the CPS—an attacker can be located in another country, within WiFi range, or possibly have direct access to the system.

The authors also introduced the multilevel framework of metrics that is aimed to correlate low level events with high level events in order to determine the attacker profile. The limitation of the approach is that it does not establish techniques and methods linking low level events with high level events. For example, the financial support metric (which can take values of low, medium, or high) expresses what budget the attacker has in order to perform an attack. However, it is not clear how the budget can be calculated on the basis of the security events registered in the system.

To conclude, modern monitoring tools and data analysis systems give new possibilities in the area of the attacker's profile construction and prediction based on the traces that the attacker leaves in the system. We argue that an approach to attack forecasting that uses relations between features in the raw security related data, attacker attributes that represent his/her behaviour, and attack development is promising for timely and efficiently counteracting cyberattacks. In this paper we start with reviewing studies that take into account such relations as soon as it is not considered in detail in the aforementioned surveys. We analyze the latest research in this area, existing challenges, and possible solutions, and conclude with a general description of the approach that can be used for forecasting attacker's goals.

Thus, the main contribution of this paper is as follows:

-   Comparative analysis and classification of existing techniques for attackers' behaviour forecasting and used characteristics of attackers.
-   Existing challenges and solutions in the considered area.

- A common approach to attack forecasting task implementation that specifies further research steps and is the basis for the development of an attacker behaviour forecasting technique.

The paper is structured as follows. The comparative analysis of the existing approaches to the attacker's profile specification, the characteristics used to describe the attacker's profile, and the attack forecasting using it are given in Section 2. Section 3 outlines existing challenges and solutions in the considered area. Besides, a common approach to attack forecasting implementation that specifies further research steps is given in Section 3, and is the basis for the development of the attacker behaviour forecasting technique. The paper ends with the conclusion and future work prospects.

2. The Comparative Analysis of the Approaches to the Attacker's Profile Specification and Attack Forecasting

The review of the existing approaches to the attacker's profile definition and attack forecasting showed that it is possible to highlight two general approaches:

(1)　the results of the attack prediction depend strongly on the attacker's model, and it is required to define the attacker's model explicitly;
(2)　the attack forecasting is based on data analysis without explicit attacker's model specification, and the attacker's behaviour is constructed implicitly on the basis of the sequence of the security events.

The first group of approaches consists of the techniques based on attack graph analysis [17–28], hidden Markov model [29–34], and fuzzy logic [35–37].

The second group of approaches consists of techniques that implement attack attribution using machine learning techniques including neural networks, statistics, and some others [38–41].

In the subsections below, these approaches are given more in detail. The summarized information on these techniques, their advantages, and their limitations is given in Table 1.

It should be noted that different researchers use not only different techniques to specify the attacker's profile, but different concepts and terms to describe attacker's behaviour, for example, "threat model", "attacker's profile", and "attacker's behaviour".

Table 1. Techniques for attacker profile specification and its application for attack forecasting. HMM, hidden Markov model; RNN, recurrent neural network; CPS, cyber-physical system.

Method	Related Research	Datasets and Features	Attacker Classification	Characteristics	Description and Advantages	Limitations
Attack graph analysis based on the analysis of network topology, software and hardware configuration, relationships between users and services, and vulnerabilities	Schneier, B., 1999 [17]; Ingols, K. et al., 2009 [19]; Kheir, N. et al., 2010 [20]; Kotenko, I. and Stepashkin, M., 2006 [21]; GhasemiGol, M. et al., 2016 [25]; Wang, L. et al., 2008 [26]; Kotenko, I. and Doynikova, E., 2018 [27]	–	• internal attacker with low skills • internal attacker with medium skills • internal attacker with high skills • external attacker with low skills • external attacker with medium skills • external attacker with high skills	• attacker skills • location	• uses a list of vulnerabilities that could be exploited by the given attacker • shows every path that an attacker can use to gain privileges • the path to be selected is determined by the attacker's skills, as well as goals and motivation	• expert knowledge to define probabilities of the next attack action selection, attacker skills and location • used attacker's model utilizes in major cases only two dimensions—skills that could be defined explicitly or implicitly, and his/her location • definition of the probabilities is a complicated process and requires great expertise of the security administrator
HMM-based approach	Rashid, T. et al., 2016 [29]; Bar, A. et al., 2016 [30]; Deshmukh, S. et al., 2019 [31]; Jhawar, R. et al., 2016 [33]	events generated by honeypots	–	–	• allows modeling normal behaviour • allows detecting insider threat • links different types of events in one model that is able to reveal trends in attack implementation and is able to detect abnormal attack sequences	• the result strongly depends on the input dataset and the distribution of the events. • does not explicitly use the attacker's model • the skills of the attackers as well as motivation, available tools, and financial support are not considered
	Katipally, R. et al., 2011 [34]	network traffic with emulated attacks	• criminal groups • insiders • terrorists • hackers • phishers • nations • spyware/malware authors • bot-net operators	• goals • intension • level of expertise	• allows modeling normal and abnormal behaviour	• the attackers' profiles are used to generate different attacks in training set, therefore, it is not used for attack prediction • the result strongly depends on the input dataset and the distribution of the events

Table 1. *Cont.*

Method	Related Research	Datasets and Features	Attacker Classification	Characteristics	Description and Advantages	Limitations
Fuzzy inference-based approach	Çayirci, E., Rong, C., 2009 [42]	• NSL-KDD CUP1999 [42,43] • CAIDA DDoS Attack 2007 Dataset [44] • qualitative attributes of the events	-	-	• deals with uncertainty existing in intrusion detection domain • allows constructing fuzzy profiles of the user behaviour or anomalous activity • fuzzification process effectively smoothens the abrupt break of normal activity and intrusion • combination of machine learning and fuzzy logic	• focus on the intrusion detection • average accuracy is 96.54% [43], classical machine learning techniques outperform approaches based on fuzzy inference • the combination with different advanced machine learning techniques such as clustering and association rule mining allows one to enhance the accuracy
	Pricop, E. and Mihalache, S.F., 2015 [35]. Mallikarjunan, K.N. et al., 2018 [36]	high level abstract variables	6–9 profiles, e.g., • script kiddie • hacker • disgruntled employee • terrorists • industrial spy • cyber warrior	No unified set of attributes to define attacker's profile. example: • resource • skills • motivation	• the main goal is the risk analysis • there is an attempt to link malicious activities to the attacker's profiles • allows describing such fuzzy parameters as motivation or knowledge to determine the attacker's profile	• there is no link with low level events generated by security sensors • proof of concept • no unified approach to define characteristics describing attacker's profile • no unified set of attacker's profiles
Fuzzy inference and attack graphs	Orojloo and Abdollahi Azgomi, 2016 [46]	high level abstract data and use case		• skills • knowledge • access (location) • interaction	• the attacker's profile is not used explicitly; however, the approach considers the characteristics of the attacker • the basis is the attack graph • the probability of the transition between vertexes is calculated with consideration of the uncertainty in data	• strongly depends on the expertise of the cyber security specialist • proof of concept
Fuzzy inference based on statistical user profiles	Kudłacik, P. et al., 2016 [37]	qualitative attributes of the log events such as keyboard keys' sequences, characteristic data sequences retrieved from pointing device, chosen options, and so on.	-	-	• allows constructing a fuzzy user profile on the basis of the statistical profiles • low computational complexity • link low level events to users' behaviour profile	• limited with detection of abnormal user's behaviour

Table 1. *Cont.*

Method	Related Research	Datasets and Features	Attacker Classification	Characteristics	Description and Advantages	Limitations
Attack attributing	Rid, T. and Buchanan, B., 2015 [38]	behavioural indicators, including atomic indicators (IP addresses, email addresses, domain names, and small pieces of text) and computed indicators ('hash')	-	-	behavioural indicators allow pointing a specific adversary who has employed similar behaviours in the past	• will be useful only in the case of correct synthesis of information flows from the technical to the operational and strategic layers
Attribution of honeypot data	Fraunholz, D. et al., 2017 [40]	• source IP address • operating system • user–agent (protocol) • cookies	• guest • external employee • internal employee • activists • state-sponsored • ethical hacker • criminals • cracker • hobby hacker	• skill • resources • motivation • intention	attempt to link raw data and high-level metrics	• though the method is introduced for IT-Security in Industry 4.0, the specific features of CPS are not considered • techniques for calculation of specific metrics require further development
RNN	Perry, I. et al., 2018 [41]	• destination port • alert signature • alert category • alert severity • proto • source port • host	–	–	• allows predicting cyber attack behaviour • accuracy of 55% for teams classification and 80% for the next alerts prediction	• depends on data sets • specific classes of attackers are not considered

2.1. Attacker Behaviour Prediction Based on Attack Graphs

The construction and application of attack graphs for attack modeling and prediction is one of the most widely used approaches. First proposed in [17], this concept was developed in many other research papers [18–28]. In the general case, an attack graph is a set of linked nodes that represents the attacker's aims and actions. The construction of the attack graph is usually based on analysis of the network topology, vulnerability analysis, and software and hardware configuration analysis, and as the result, it shows dependencies between vulnerabilities and the overall security state of the target network.

In major cases, the attacker's model is defined via two important characteristics—his/her skills and location. For example, in the literature [19,21,27], these attributes are used to implement attack reachability analysis depending on the location (internal or external) and skills of the attacker (low, medium, or high). In fact, the level of the attacker's skills defines a list of vulnerabilities that could be exploited by the given attacker. In [27], the attacker's skills are correlated with meanings of "attacker skills" or "knowledge required" parameters of the attack patterns defined in Common Attack Pattern Enumeration and Classification (https://capec.mitre.org/) database and weaknesses from Common Weakness Enumeration (https://cwe.mitre.org/) database. This allows authors to link existing vulnerabilities to high-level malefactor activity such as "host discovery", "active operating system (OS) fingerprinting", and so on.

Wang et al. 2008 [26] assigned to each malefactor action a score that reflected the probability of its implementation. This score implicitly defines the attacker's skills, and in the approach, it was determined on the basis of the expert's knowledge regarding the vulnerability being exploited. Kheir et al. [20] enhanced the attack graph model by adding the service-dependency graph, which presents a network model for the relationships between users and services, showing how they perform their activities using the available services in order to increase the efficiency of the attack modeling.

In [25], the authors introduced the concept of the uncertainty-aware attack graph, which is used to handle the uncertainty of attack probability. This uncertainty appears owing to the measuring probability of vulnerability exploitation. In fact, it is difficult to find the precise probabilities for all attack graph nodes, and the authors suggest assigning the node probability in the form of interval values or constraints. However, both probability intervals and constraints are set by the experts. For example, the constrain may be described as follows [25]: "The probability of attack on workstation is greater than the probability of attack on webserver plus 0.05".

The experiments showed that the introduction of the uncertainty to the attack graph modeling and forecasting, on one hand, adds extra flexibility to the security administrator and may significantly reduce the attack graph, resulting in its better comprehensiveness. On the other hand, the definition of the probabilities and constraints is a complicated process and requires great expertise of the security administrator.

A set of European research projects devoted to the attacker's behaviour prediction as well as risk assessment utilized the approach based on analysis of the attack graphs, including TREsPASS (https://cordis.europa.eu/project/id/318003) (Technology-supported Risk Estimation by Predictive Assessment of Socio-technical Security) and MASSIF (MAnagement of Security information and events in Service InFrastructures) [47].

The TREsPASS project is interesting in that, when constructing an attack graph, the authors consider not only software exploits and configuration weaknesses, but also physical entities that could be used to gain access to the information resources. As the result, they developed the special attack navigator map tool, which allows uniting computer network entities and physical objects of the critical infrastructure, highlighting the fact that the attack may be implemented on both the networking level and the level of the physical objects. The forecasting of the malefactor actions considers the attacker's profiles presented in [48]. These profiles, known as threat agents, are based on eight attributes: intent, access, outcome, limits, resource, skill level, objective, and visibility.

To summarize, it is possible to say that attack graphs show every possible path that an attacker can use to gain further privileges—the path to be selected is determined by the attacker's skills as well as goals and motivation. In the general case, the attack graph complexity is $O(scn^2)$, for n machines in the attack graph, where s is the average number of exploits per machine and c is the average number of security conditions per machine. The survey of the graph-based techniques showed that the used attacker's model utilizes, in major cases, only two dimensions of the attacker's model—skills that could be defined explicitly or implicitly, and his/her location. Obviously, understanding the attacker's motivation and goal could significantly reduce the complexity of the attack graph and, as a result, increase the efficiency of the attack forecasting.

2.2. Attacker Behaviour Prediction Based on Hidden Markov Model

The Markov-based methods are very close to the attack tree models. In general, they are constructed on the basis of system states, and transitions between them, caused by events. Each transition is characterized by a probability that is independent of the past, and depends only on the two states involved—the behaviour of a process at a given point in time depends only on the state of the process at a previous point in time. The hidden Markov models (HMMs) for modeling normal behaviour to detect cyber attacks were first proposed in [29]. The authors used them to describe normal behaviour of the users as a sequence of the events and then applied them to detect insider threat. Since then, a significant amount of research has been done to enhance the HMM and its learning algorithm for detecting and predicting cyber attacks [30,31,33,34]. They vary in structure of HMM, used datasets, and particular tasks solved.

For example, in [30], the authors used HMMs to model and predict attack propagation based on data from different types of honeypots. In the research, they used data from the following families of honeypots:

- Glastopf (https://github.com/mushorg/glastopf)—a honeypot that emulates vulnerabilities that are relevant to web applications;
- Kippo (https://github.com/desaster/kippo)—a medium-interaction SSH honeypot;
- Honeytrap (https://www.honeynet.org/projects/active/honeytrap/)—a low-interaction honeypot that aims at collecting malware in an automated way;
- Dinoaea (https://www.div0.sg/single-post/dionaea-malware-honeypot)—malware capturing a honeypot that emulates several well-known protocols.

Thus, the authors managed to link different types of events in one model that is able to reveal trends in attack implementation and is able to detect abnormal attack sequences.

In [31], the authors applied a set of HMMs named as the fusion hidden Markov model. They construct k HMMs on k different low-correlated partitions of data and make a prediction using a nonlinear weight function. The latter is implemented by a neural network that is trained on the predictions of HMMs to the next state output. The application of k HMMs defines rather strict requirements to the HMMs; they have to be diverse and low correlated. To fulfill this requirement, the authors use a dissimilarity function to divide data into k different subsets, such that each subset contains a particular temporal pattern of the data. The input data are the real attack logs collected by the Cowrie honeypot [32], which is a medium-interaction SSH and telnet honeypot. The authors divided them into 19 groups corresponding to different activities, and these groups were modeled as states of the HMMs.

In [33], the continuous time Markov chain is used to make a prediction of the attack propagation.

It is clearly seen that this group of approaches does not use the attacker's model explicitly. The result of the prediction by the HMM strongly depends on the input dataset and the distribution of the events. The prediction of the attack goal is done on the basis of the most probable transition for the current system state, that is, the most frequently met sequence of the events. The skills of the attackers as well as motivation, available tools, and financial support are not considered.

In [34], the authors specify the attacker behaviour based on their goals, intention, and level of expertise, and outlined eight profiles of the attackers such as criminal groups, insiders, terrorists, hackers, phishers, nations, spyware/malware authors, and bot-net operators. However, the definition of the HMM presented in their approach did not consider the attacker's profile. The HMM is described as follows:

$$\lambda = (A, B, \pi, N),$$

where N corresponds to five different types of malicious behaviour (scanning, enumeration, access attempt, malware attempt, exploitation by denial of service), where

π is the state probabilities,

A is the transition probabilities, and

B is the observation probabilities.

Interestingly, the authors used the attacker's profiles for generating different training sets containing five types of malicious behaviour.

2.3. Attacker Behaviour Pattern Discovery Using Fuzzy Inference

The benefits of the fuzzy logic approaches consist in their ability to operate with uncertainty. We consider several works devoted to the intrusion detection based on fuzzy logic [45,49–51]. In major cases, fuzzy logic is applied to produce some averaged description of the parameters used to describe either normal or malicious activities. For example, in [50], the fuzzification process is applied to the metrics describing TCP service channel between two IP end-points—count, uniqueness, and variance. The authors defined five fuzzy sets for each metric: LOW, MEDIUM-LOW, MEDIUM, MEDIUM-HIGH, and HIGH, and defined the fuzzy set distributions using historical data. The authors applied fuzzy rules constructed as a combination of these parameters to determine the type of malicious activity, such as port scanning. In [51], fuzzy rules are constructed based on the results obtained by association rule mining. In [45], the authors applied leader-based k-means clustering to preprocess data before application of the fuzzification process. Thus, the existing approaches differ in preprocessing steps and data attributes to construct fuzzy rules for classifying the types of the malicious activities.

In [37], the authors solve the problem of constructing profiles of the normal user behaviour based on the analysis of the log events such as keyboard keys' sequences, characteristic data sequences retrieved from pointing device, chosen options, requested network resources, and so on. They apply fuzzy logic to the qualitative attributes of these events to describe a set of fuzzy profiles and identify masqueraded attacks.

In [46], an approach to combining attack graphs and fuzzy logic to predict attacker's behaviour was suggested. The attack graph is constructed in a traditional manner as a sequence of possible malefactor steps. Four parameters characterizing the attacker are assigned to each step: "the required knowledge for performing the attack action; (ii) the required access for conducting the attack action (the attack step may need physical access or it can be performed remotely); (iii) the required user interaction level for successful preformation of the attack (such as social engineering attacks against employees or the attacks targeting human-machine interface operators); and (iv) the required skill for conducting the attack" [46]. These parameters take the following values: low importance, moderate importance, importance, high importance, and very high importance. The fuzzy sets are described by triangular function. The complexity of the attack step depends on the values of these four variables. Apart of the assessment of the complexity of each attack step, the authors rate the alternatives existing for each attack step. This rating reflects the attractiveness of each step for the attacker and is evaluated on the basis of the expert's assessments. It is also a fuzzy variable that takes the following values: very low, moderate, high, and very high. To make a prediction of the attack deployment, the authors apply the Technique for Order of Preference by Similarity to Ideal Solution (TOPSIS) approach, which is a multi-criteria decision making method suggested for fuzzy environment [52]. It allows the analyst to compare alternatives described by fuzzy variables. The general scheme of the approach is given in Figure 1 [46].

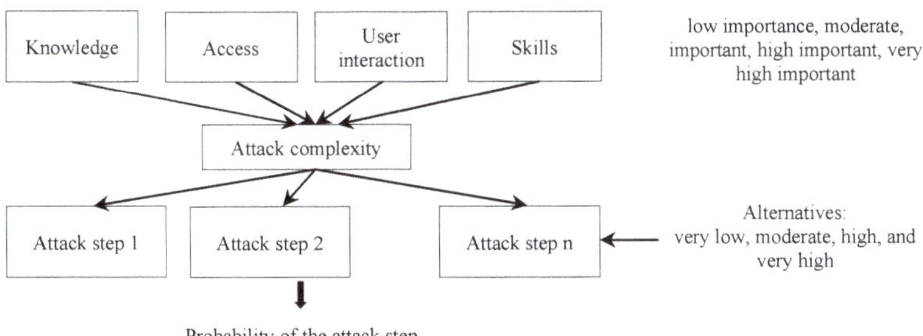

Figure 1. The overall scheme of the approach based on the combination of the attack graph and fuzzy logic.

Pricop and Mihalache [35] apply a fuzzy approach to model the impact of cyber attacks. Like in [46], the authors describe the attacker's profile as a combination of the following three parameters: knowledge, technical resources, and motivation—that is, a function of three inputs and one output. They define six types of attackers, as follows: script kiddie, hacker, disgruntled employee, terrorists, industrial spy, and cyber warrior. The script kiddie is an inexperienced and unskilled attacker that uses known exploits, and whose motivation is usually to get reputation, while the cyber warrior has the highest levels of knowledge, resources, and motivation. The cyber warrior is the most dangerous type of attacker, targeting the critical infrastructure.

The variables describing the attacker profile are linguistic variables that take values from fuzzy sets—very small, small, medium, big, and very big—which are presented by triangular curves. The highest score is assigned to the industrial spy; the cyber warrior, terrorist, and disgruntled employee have a medium score; the hacker's score is small; and the script kiddie has a very small score.

The attacker's profile, that is, the score [35], is used then to estimate the attack success rate. The impact of the attack is also a fuzzy function of four linguistic variables: the attacker profile (score), protection level, vulnerabilities, and restore cost. In the approach, these variables are described by a membership function of triangular form, defined for three fuzzy values—small, medium, and big. The attack success rate allows the analyst to understand how these parameters influence the overall security state of the information system.

In [36], the authors try to link attack steps to produce the attacker's profile. They developed a fuzzy inference system that takes as input the following linguistic variables: scanning/reconnaissance, enumeration, exploit by access attempt, exploit by denial of service, exploit by malware attempt, and output the attacker category. The possible attacker's categories are as follows: criminals, insiders, terrorists, hackers, phishers, nations, spyware/malware authors, bot-net operators, and amateur/script kids.

The linguistic variables used to determine the attacker's category may take the following fuzzy values: none, low, and high, which are described by a triangular form.

Thus, it is possible to conclude that there are two broad groups of approaches based on fuzzy logic to predict the attacker's behaviour.

The first group of techniques uses fuzzy inference to detect the type of the malicious activity, and the fuzzy rules describe generalized (fuzzy) dependencies between security event attributes. It is worth noticing that, in major cases, the authors apply fuzzy inference to detect attacks that have rather specific characteristics, such DoS attacks and port scanning. It could be explained that the most widely used data sets are NSL-KDD CUP 1999 and CAIDA UCSD "DDoS Attack 2007". These datasets do not contain complicated long-term attacks. They also do not consider attacks targeting IoT-based infrastructures, cyber-physical systems, "smart" homes, and so on.

The second group of techniques mostly focuses on risk assessment and uses the attacker's profile explicitly as an input variable that defines the success rate of the attack. The advantage of the application of fuzzy logic is the ability to describe such fuzzy parameters as the motivation or knowledge of the malefactor. However, the major limitation of this group is the inability to link low level events to the attributes used to characterize the malefactor profile. The possible solution is to implement consequent mapping of low level events to middle level activities, and then determine the high level attributes of the attacker such as skills, resources, and motivation.

2.4. Attributing Cyber Attacks

In [38], the concept of attack attributing is used, that is, the determination of attack author, based on behavioural indicators. Behavioural indicators are combinations of actions and other indicators of malicious activity. These indicators can be atomic and computed. Atomic indicators are discrete pieces of data that cannot be broken down into their components without losing their forensic value. Atomic indicators include IP addresses, email addresses, domain names, and small pieces of text. Computed indicators are similarly discrete pieces of data, but they involve some element of computation. An example is a 'hash', a unique signature derived from input data, for instance, a password or a program. Hashes of programs running on their network's computers may match hashes of programs known to be malicious.

In some cases, behavioural indicators point to a specific adversary who has employed similar behaviours in the past. It might be repeated social engineering attempts of a specific style via email against low-level employees to gain a foothold in the network, followed by unauthorized remote desktop connections to other computers on the network delivering specific malware.

The authors outline that, though details are critical for attacker attributing, they will be useful only in the case of correct synthesis of information flows from the technical to the operational and strategic layers.

In [39], the authors build a cyber attacker model profile (CAMP) that can be used to characterize and predict cyber attacks. The authors define two types of variables used—dependable and independent. They denote the frequency and distribution of attacks as well as money earned from cybercrime as dependable variables (DVs), while unemployment rate, level of education, and corruption are independent variables. The authors constructed the attack prediction model linking both types of variables and showed how much variation in the DVs they can explain for given values of independent variables.

In [40], the attribution of honeypot data is considered. The authors define an attacker via a unique tuple (source IP address, operating system, user–agent (protocol), {cookies}). They assumed that the knowledge of the operating system, user agent, and set of cookies allows more accurate classification than the source IP address only. Honeypot data (HD) are used to calculate skill, resources, motivation, and intention. Further, they integrate skill (S) and resources (R) into the capability rating, and motivation (M) and intention (I) into the threat rating. Their combination is used to calculate the total threat score. S, R, M, and I are determined by weighted accumulation of all affecting features f_i:

$$V = \sum_{i=1}^{n} a_i f_i,$$

where n is the total number of features f_i;

a_i is the weight for the i-th feature, $\sum_{i=1}^{n} a_i = 1$.

The features f_i are derived from the considered observed HD features v_i and get values of $\{0, \ldots, \gamma\} \in \mathbb{Q}$. The maximal value of S, R, M, and I is γ. The dimension and boundaries for v_i vary between the parameter and sensor resolution. The part of sample feature set provided by the authors is represented in Table 2.

Table 2. Part of sample feature set for attackers' classification suggested in [40].

Origin									Temporal	
Port		IP address		User agent	URL	Domain	E-mail	User ID	OS	Inter-Arrival Time	
Protocol	Service	Autonomous system	Country							Standard deviation	Average

For example, to calculate R, the assumption can be made that fast inter-arrival times are related to a higher degree of automation (higher attackers' resources). The motivation attribute can be estimated by the time and effort an attacker invests into a particular attack. Quantifying an attacker's intention is the most complex task. The authors define intention as the degree or potential of attacker's maliciousness.

The authors use the following classes of attackers [40]: guest (G), external employee (E), internal employee (I), activists (A), state-sponsored (N), ethical hacker (W), criminals (O), cracker (C), and hobby hacker (H).

The values for different classes are calculated using $V \in \{S,R,I,M\}$, which are ordered as $V_{ci} < V_{cj} \ldots < V_{cn}, \forall c \in C$, and then transformed to $\{0,\ldots,\gamma\} \in \mathbb{Q}$, by assigning 1 to the first class and iterating over all classes while incrementing the value by 1 for each less-than operator. Then, all values are normalized with $\gamma = 10$. In Figure 2, the heat map proposed by the authors to represent attackers' classes is provided (the capitalized abbreviation marks the appropriate class).

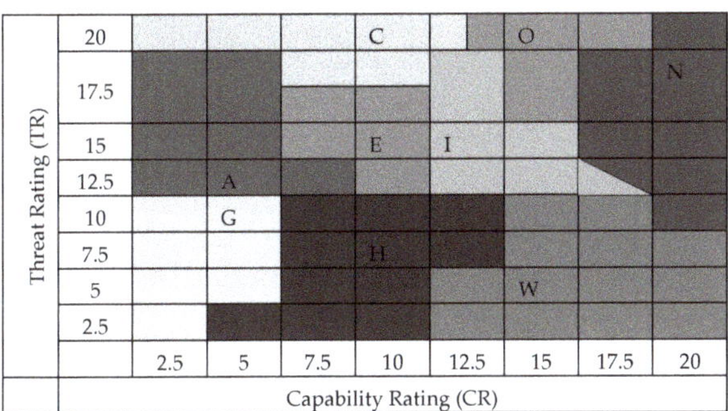

Figure 2. Attackers' classes [40].

Though the method is introduced for IT-Security in Industry 4.0, nonetheless, the specific features of CPS are not considered.

In [41], the authors propose the method for predicting the behaviour of cyberattacks using recurrent neural networks (RNNs). They use the dataset obtained from the 2017 Collegiate Penetration Testing Competition (CPTC) to obtain long-short-term-memory (LSTM) models. The dataset includes Suricata alerts obtained, while ten student teams attempted to penetrate the virtualized network and exploit vulnerabilities. The authors trained two sets of models: the first set determines the team that caused the alert, and the second predicts the second alert. The used features are as follows: destination port, alert signature, alert category, alert severity, proto, source port, and host. The authors achieved accuracy of 55% for teams classification and 80% for the next alerts prediction.

Finally, while the last works on the attacker behaviour forecasting using machine learning make attempts to overcome the challenge of linking raw data with valuable attacker metrics, the feature set is still not specified, the set of metrics that forms the attacker profile is not unified, the techniques of metrics calculation on the basis of the extracted features should be enhanced, and the training dataset problem still exists.

3. Challenges, Possible Solutions, and Common Approaches

The analysis conducted allowed us to conclude on the main challenges existing in attack goal forecasting to this moment:

1. Lack of uniformity in the classification of attackers, distinguished metrics, and attributes, as well as the definition of the same classes and metrics.
2. The gap between the raw data (such as network traffic and events logs), attacker profile, and forecasting of the attacker behaviour, as well as the methods for the determination of relationships between them.
3. Lack of datasets suitable for research of the relationships between attacker steps and his/her goals.
4. Absence of the research that demonstrates if there is an influence of the attacker profiling and attributing on the attack forecasting.

The lack of uniformity indicates insufficient elaboration of the problem under research. Besides, it prevents efficient countermeasure selection for different classes of attackers, understanding of current research state, comparative quantitative analysis of the various developed techniques, and elaboration of the existing results. An attempt to overcome this challenge was made in [9]. However, the authors do not describe how to link low level events with high level events, that is, they did not proposed a solution to the second problem.

Considering the second challenge, an attempt to link low level events with security metrics was made in the Structured Threat Information Expression (STIX) project [53]. Structured Threat Information Expression (STIX) is a structured language for specification of various threats and automated analysis. The idea behind the development of this language is to link low level events with high level concepts. The following components of the language are specified [54]: observables; indicators (observation patterns and their meanings); incidents (attack actions instances); adversary tactics, techniques, and procedures (methods that are used by an attacker, including attack patterns, malware, exploits, and so on); exploit targets (e.g., vulnerabilities, weaknesses, and configurations); courses of action (response actions to prevent an attack); campaigns (sets of incidents and/or TTPs with a single goal); threat actors (attacker identification); and reports.

For each component, the set of properties is specified. For example, for threat actors, the following properties are used: name, description, aliases, roles, goals, sophistication, resource_level, primary_motivation, secondary_motivations, and personal_motivations.

All properties are of a nominal type (i.e., values are selected from a list). Thus, for threat actor labels, the possible values are as follows: activist, competitor, crime-syndicate, criminal, hacker, insider-accidental, insider-disgruntled, nation-state, sensationalist, spy, and terrorist. While for the threat actor sophistication (captures the skill level of a threat actor; ranges from "none", which describes a complete novice, to "strategic", which describes an attacker who is able to influence supply chains to introduce vulnerabilities), the values are as follows: none, minimal, intermediate, advanced, expert, innovator, and strategic. In this project, however, how to determine the values of these properties automatically from the raw data is not also described. It should be actively used by the security companies in order to reveal and then automate the process of linking low events and high level attack concepts; however, there is not much activity in this field.

The second challenge is connected with the third challenge, that is, the absence of datasets for analysis aimed at revealing existing interrelations and features characterizing attackers and their goals. The following approaches are used to overcome it:

- Use existing datasets with specific attacks' data.
- Use honeypots to generate real data.
- Use normal data and add data on attacks intentionally (use attack generators).

The first approach is used for the detection of specific types of attacks based on training using the datasets. However, the most used datasets are deprecated and do not represent the last trends in attacks or paradigm of the modern information systems.

According to the second approach, in [30], the authors used the honeypot technology. The detailed description of attack features logged and dataset description, when using the honeypot technology, is provided in [55]. The analysis is based on the following assumption: the data are grouped by session ID for considering that the attacker attempts to implement some malicious scenario in one session, that is, different session IDs are independent of individual attacker characteristics. This allows the authors to group event sequences to create a training sample by sessions. However, this approach does not consider an opportunity to use several sessions to implement a complex multistep attack by a single attacker.

The researchers usually create their own datasets and use them [56]. Unfortunately, however, in most of these approaches, all these data are not annotated by attackers, that is, their skills, knowledge, and other characteristics that form their profiles. In fact, datasets contain only attacks of different types, and there are no labeled datasets characterizing attackers' skills. This is explained by the fact that the techniques used to detect attacks analyze the event sequences, their frequencies, and attributes. Until there is no research proving that the application of the attacker attribution may enhance the efficiency of the attack detection, there will be no datasets linking raw security events with attacker's profile concepts such as attacker motivation, goal, and so on. However, having such a dataset maybe extremely useful in detecting targeted and distributed in time attacks. Unfortunately, the absence of datasets is a common problem that can be solved with their targeted generation.

Thus, we argue that there is a need in the research that demonstrates if there is an influence of the attacker profiling and attributing on the attack forecasting. Thus, the fourth challenge is one of our future research directions. However, it is necessary to overcome the first three challenges first. In particular, we are planning start with the generation of the specific dataset. We consider that the approach presented in [57] is the promising one to generate datasets for attack attribution. It is based on mixed traffic generation, including attacks and normal traffic.

To conclude, we propose the following approach to the attacker behaviour forecasting:

1. First of all, we suppose to outline possible raw data sources. There are two types of sources: structured data and unspecified data. In [58], we outlined the following open sources of structured data considering objects of information security assessments: vulnerability databases, attack patterns databases, weaknesses databases, software and hardware databases, and so on. For accurate attack forecasting in real time, it is required to add another type of source data, network traffic, and event logs (which is unspecified). From the analyzed events datasets, the most interesting is the one provided in [56]. The dataset should contain data on various attacks with different goals implemented by attackers of different classes. From our point of view, the most complete classification from those reviewed was proposed in [40]. It incorporates the following classes: guest, external employee, internal employee, activists, state-sponsored, ethical hacker, criminals, cracker, and hobby hacker.

2. Extract features from the events dataset that can characterize different classes of attackers with different goals. While there are rather detailed sets of features from the network traffic (such as source IP address, operating system, user–agent (protocol), and {cookies} in [40]), the events features should be researched in more detail. In [41], the following set is proposed: destination port, alert signature, alert category, alert severity, proto, source port, and host. We can use this as the basis for future research.

3. Then, we suppose to outline and classify high level metrics that form the attacker profile, on the basis of the following metrics, proposed in [59]: attacker skill level, attacker knowledge, tools complexity, attack steps complexity, steps success rate, trace coverage rate, and so on.

4. Then, we propose to find out structural and semantic relations between data sources, objects of the attacker behaviour forecasting subject area, and metrics (from features extracted from the raw

data to high level metrics of attackers and attacks). To implement these, we plan to extend an ontology provided in [59] and determine transitional metrics.

5. Then, we propose to use the outlined characteristics and relationships to do the following:

   a. develop algorithms for metrics calculation;
   b. train a neuro-fuzzy network for attackers' behaviour forecasting.

We state that steps 1–4 are the necessary basis for step 5, while overcoming challenges 1–4 is the basis for the successful implementation of our research task.

Thus, at this stage, we developed the common approach to forecasting attacker goals and considered the future work scope on the basis of comparative analysis of the related research and existing challenges in the area.

4. Conclusions

In the paper, we reviewed the research in the area of attacker behaviour forecasting. Compared with the close survey described in [4], our research is focused on issues of the attacker profile definition or attacker attribution and its influence on the attack forecasting process. In [9], an interesting study that highlights main challenges in the area of attacker behaviour forecasting is provided and the multilevel system of metrics is introduced. Our goal in this research, however, is to determine how to link low level events with high level events. Besides, compared with the aforementioned papers, the main goal of the research outlined in the paper is the novel approach development. Though the main goal of the research outlined in the paper is not devoted only to the state-of-the-art, it is necessary for novel approach development. In the scope of our research, we considered four classes of approaches to the attacker behaviour forecasting, including attack graph based approach, HMM, fuzzy inference, and approaches based on intelligent data processing. The analysis shows that there is a lack of formalization and systematic representation of the attacker profile and of the definition of his/her characteristics that can be used for his/her specification. From our point of view, the most promising are approaches based on intelligent data analysis, as soon as they allow linking raw data and metrics describing an attacker.

The conducted analysis allowed us to outline the key challenges in the area. On the basis of these challenges and our task, we have selected the approach to the task implementation. The proposed approach specifies our further research steps and is the basis for the technique of attacker behaviour and goals forecasting under development.

The approach incorporates the following steps: (1) outline raw data sources, both structured and unspecified; (2) extract features from the events dataset that characterize different classes of attackers with different goals; (3) outline and classify high level metrics that form attacker profile; (4) find out structural and semantic relations between data sources, objects of the attacker behaviour forecasting subject area, and metrics (from features extracted from the raw data to high level metrics of attackers and attacks); and (5) use the outlined characteristics and relationships to develop algorithms for metrics calculation, and to train neuro-fuzzy network for attackers' behaviour forecasting. Compared with the other approaches, summarized in this paper, our approach is focused on the accurate determination of relations among raw data and attacker behaviour characteristics. Each step of the proposed approach will be discussed in detail in the following research. Moreover, in the scope of our future research, we will analyze if there is the influence of the attacker profiling and attributing on the attack forecasting.

Author Contributions: Investigation, E.D. and E.N.; literature analysis, E.D. and E.N.; common approach, E.D. and E.N.; writing—original draft, E.D. and E.N.; writing—review and editing, E.D., E.N. and I.K. All authors have read and agree to the published version of the manuscript.

Funding: This research received no external funding.

References

1. Howard, J.D.; Longstaff, T.A. *A Common Language for Computer Security Incidents*; Sandia National Labs.: Albuquerque, NM, USA; Livermore, CA, USA, 1998.
2. Abomhara, M.; Køien, G.M. Cyber security and the Internet of Things: Vulnerabilities, threats, intruders and attacks. *J. Cyber Secur. Mob.* **2015**, *4*, 65–88. [CrossRef]
3. Aliyev, V. Using Honeypots to Study Skill Level of Attackers Based on the Exploited Vulnerabilities in the Network. Ph.D. Thesis, Chalmers University of technology, Göteborg, Sweden, 2010.
4. Gheyas, I.A.; Abdallah, A.E. Detection and prediction of insider threats to cyber security: A systematic literature review and metaanalysis. *Big Data Anal.* **2016**, *1*, 6. [CrossRef]
5. Husák, M.; Komárková, J.; Bou-Harb, E.; Čeleda, P. Survey of Attack Projection, Prediction, and Forecasting in Cyber Security. *IEEE Commun. Surv. Tutor.* **2019**, *21*, 640–660. [CrossRef]
6. Yang, S.J.; Du, H.; Holsopple, J.; Sudit, M. Attack Projection. In *Cyber Defense and Situational Awareness*; Springer: Cham, Switzerland, 2014; pp. 239–261.
7. Ahmed, A.A.; Zaman, N.A.K. Attack intention recognition: A review. *IJ Netw. Secur.* **2017**, 244–250.
8. Abdlhamed, M.; Kifayat, K.; Shi, Q.; Hurst, W. Intrusion Prediction Systems. In *Information Fusion for Cyber-Security Analytics*; Springer: Cham, Switzerland, 2017; pp. 155–174.
9. Rocchetto, M.; Tippenhauer, N.O. On Attacker Models and Profiles for Cyber-Physical Systems. In *Lecture Notes in Computer Science, Proceedings of the ESORICS, 2016*; Askoxylakis, I., Ioannidis, S., Katsikas, S., Meadows, C., Eds.; Springer: Cham, Switzerland, 2016.
10. Corman, J.; Etue, D. Adversary ROI: Evaluating Security from the Threat Actor's Perspective. In Proceedings of the RSA Conference Europe 2012, San Francisco, CA, USA, 9–11 October 2012.
11. Heckman, R.M. Attacker Classification to Aid Targeting Critical Systems for Threat Modelling and Security Review. 2005. Available online: www.rockyh.net/papers/AttackerClassification.pdf (accessed on 22 January 2020).
12. Cardenas, A.A.; Amin, S.M.; Sinopoli, B.; Giani, A.; Perrig, A.; Sastry, S.S. Challenges for Securing Cyber Physical Systems. In *Workshop on Future Directions in Cyber-physical Systems Security*; DHS: Newark, NJ, USA, 2009.
13. LeMay, E.; Ford, M.D.; Keefe, K.; Sanders, W.H.; Muehrcke, C. Model-Based Security Metrics Using Adversary View Security Evaluation (ADVISE). In Proceedings of the 2011 Eighth International Conference on Quantitative Evaluation of Systems, Aachen, Germany, 5–8 September 2011.
14. Cardenas, A.A.; Roosta, T.; Sastry, S. Rethinking security properties, threat models, and the design space in sensor networks: A case study in SCADA systems. *Ad Hoc Netw.* **2009**, *7*, 1434–1447. [CrossRef]
15. Matusitz, J. Cyberterrorism: Postmodern state of chaos. *Inf. Secur. J.* **2008**, *17*, 179–187. [CrossRef]
16. Denning, D.E. Activism, hacktivism, and cyberterrorism: The internet as a tool for influencing foreign policy. *Netw. Netwars Future Terror Crime Militancy* **2001**, *239*, 288.
17. Schneier, B. Attack Trees—Modeling Security Threats. *Dr. Dobb's J.* **1999**, *24*, 12.
18. Hariri, S.; Qu, G.; Dharmagadda, T.; Ramkishore, M.; Raghavendra, C.S. Impact Analysis of Faults and Attacks in Large-Scale Networks. *IEEE Secur. Priv.* **2003**, *1*, 49–54. [CrossRef]
19. Ingols, K.; Chu, M.; Lippmann, R.; Webster, S.; Boyer, S. Modeling Modern Network Attacks and Countermeasures Using Attack Graphs. In Proceedings of the 2009 Annual Computer Security Applications Conference (ACSAC'09), Honolulu, HI, USA, 7–11 December 2009.
20. Kheir, N.; Cuppens-Boulahia, N.; Cuppens, F.; Debar, H. A Service Dependency Model for Cost-Sensitive Intrusion Response. In Proceedings of the 15th European Symposium on Research in Computer Security (ESORICS), Athens, Greece, 20–22 September 2010.
21. Kotenko, I.; Stepashkin, M. Attack Graph based Evaluation of Network Security. *Lect. Notes Comput. Sci.* **2006**, *4237*, 216–227.
22. Kotenko, I.; Stepashkin, M.; Doynikova, E. Security Analysis of Computer-aided Systems Taking into Account Social Engineering Attacks. In Proceedings of the 19th Euromicro International Conference on Parallel, Distributed and Network-Based Processing (PDP 2011), Los Alamitos, CA, USA, 9–11 February 2011; pp. 611–618.
23. Noel, S.; Jajodia, S.; O'Berry, B.; Jacobs, M. Efficient minimum-cost network hardening via exploit dependency graphs. In Proceedings of the 19th Annual Computer Security Applications Conference (ACSAC'03), Las Vegas, NV, USA, 8–12 December 2003.

24. Wang, L.; Jajodia, S.; Singhal, A.; Noel, S. k-Zero Day Safety: Measuring the Security Risk of Networks against Unknown Attacks. In Proceedings of the 15th European Conference on Research in Computer Security; Springer: Berlin/Heidelberg, Germany, 2010; pp. 573–587.
25. GhasemiGol, M.; Ghaemi-Bafghi, A.; Takabi, H. A comprehensive approach for network attack forecasting. *Comput. Secur.* **2016**, *58*, 83–105. [CrossRef]
26. Wang, L.; Islam, T.; Long, T.; Singhal, A.; Jajodia, S. An Attack Graph-Based Probabilistic Security Metric. In *Lecture Notes in Computer Science 5094, Proceedings of the Data and Applications Security XXII (DBSec 2008)*; Atluri, V., Ed.; Springer: Berlin/Heidelberg, Germany, 2008.
27. Kotenko, I.; Doynikova, E. Improvement of attack graphs for cybersecurity monitoring: Handling of inaccuracies, processing of cycles, mapping of incidents and automatic countermeasure selection. *SPIIRAS Proc.* **2018**, *57*, 211–240.
28. An, S.; Eom, T.; Park, J.S.; Hong, J.B.; Nhlabatsi, A.; Fetais, N.; Khan, K.M.; Kim, D.S. CloudSafe: A Tool for an Automated Security Analysis for Cloud Computing. Available online: https://arxiv.org/abs/1903.04271v1 (accessed on 25 January 2020).
29. Rashid, T.; Agrafiotis, I.; Nurse, J.R.C. A New Take on Detecting Insider Threats: Exploring the Use of Hidden Markov Mode. In Proceedings of the 8th ACM CCS International Workshop on Managing Insider Security Threats, Vienna, Austria, 28 October 2016; ACM: New York, NY, USA, 2016; pp. 47–56.
30. Bar, A.; Shapira, B.; Rokach, L.; Unger, M. Identifying Attack Propagation Patterns in Honeypots Using Markov Chains Modeling and Complex Networks Analysis. In Proceedings of the IEEE International Conference on Software Science, Technology and Engineering (SWSTE), Beer Sheva, Israel, 23–24 June 2016; pp. 28–36.
31. Deshmukh, S.; Rade, R.; Kazi, F. Attacker Behaviour Profiling Using Stochastic Ensemble of Hidden Markov Models. 2019. Available online: https://arxiv.org/abs/1905.11824 (accessed on 25 January 2020).
32. Oosterhof, G.M. Cowrie—Medium-Interaction Honeypot. Available online: https://github.com/micheloosterhof/cowrie (accessed on 25 January 2020).
33. Jhawar, R.; Lounis, K.; Mauw, S. A Stochastic Framework for Quantitative Analysis of Attack-Defense Trees. In *Lecture Notes in Computer Science 9871, Proceedings of the Security and Trust Management (STM 2016)*; Barthe, G., Markatos, E., Samarati, P., Eds.; Springer: Cham, Switzerland, 2016.
34. Katipally, R.; Yang, L.; Liu, A. Attacker Behavior Analysis in Multi-stage Attack Detection System. In Proceedings of the Cyber Security and Information Intelligence Research (CSIIRW'11), Oak Ridge, TN, USA, 12–14 October 2011; ACM: New York, NY, USA, 2011. Available online: https://www.utc.edu/center-academic-excellence-cyber-defense/pdfs/paper-csiirw-2011-attacker-behavior.pdf (accessed on 25 January 2020).
35. Pricop, E.; Mihalache, S.F. Fuzzy approach on modelling cyber attacks patterns on data transfer in industrial control systems. In Proceedings of the 7th International Conference on Electronics, Computers and Artificial Intelligence (ECAI 2015), Bucharest, Romania, 25–27 June 2015.
36. Mallikarjunan, K.N.; Shalinie, S.M.; Preetha, G. Real Time Attacker Behavior Pattern Discovery and Profiling Using Fuzzy Rules. *J. Internet Technol.* **2018**, *19*, 1567–1575.
37. Kudłacik, P.; Porwik, P.; Wesołowski, T. Fuzzy approach for intrusion detection based on user's commands. *Soft Comput.* **2016**, *20*, 10–16. [CrossRef]
38. Rid, T.; Buchanan, B. Attributing Cyber Attacks. *J. Strateg. Stud.* **2015**, *38*, 4–37. [CrossRef]
39. Watters, P.A.; McCombie, S.; Layton, R.; Pieprzyk, J. Characterising and Predicting Cyber Attacks Using the Cyber Attacker Model Profile (CAMP). *J. Money Laund. Control* **2012**, *15*, 430–441. [CrossRef]
40. Fraunholz, D.; Krohmer, D.; Duque Antón, S.; Schotten, H.D. YAAS—On the Attribution of Honeypot Data. *Int. J. Cyber Situat. Aware.* **2017**, *2*, 31–48. [CrossRef]
41. Perry, I.; Li, L.; Sweet, C.; Su, S.H.; Cheng, F.-Y.; Yang, S.J.; Okutan, A. Differentiating and Predicting Cyberattack Behaviors Using LSTM. In Proceedings of the 2018 IEEE Conference on Dependable and Secure Computing (DSC), Kaohsiung, Taiwan, 10–13 December 2018.
42. Çayirci, E.; Rong, C. *Security in Wireless Ad Hoc and Sensor Networks*; John Wiley & Sons: Hoboken, NJ, USA, 2009.
43. Kdd Cup 1999 Data. UCI KDD Archive. 1999. Available online: http://kdd.ics.uci.edu/databases/kddcup99/kddcup99.html (accessed on 25 January 2020).

44. The CAIDA DDoS Attack 2007 Dataset. 2007. Available online: http://www.caida.org/data/passive/ddos-20070804_dataset.xml (accessed on 25 January 2020).

45. Shyla, S.; Sujatha, S. Cloud Security: LKM and Optimal Fuzzy System for Intrusion Detection in Cloud Environment. *J. Intell. Syst.* **2019**, *29*, 1626–1642. [CrossRef]

46. Orojloo, H.; Abdollahi Azgomi, M. Predicting the behavior of attackers and the consequences of attacks against cyber-physical systems. *Secur. Commun. Netw.* **2016**, *9*, 6111–6136. [CrossRef]

47. MASSIF FP7 Project. MASSIF Architecture. 2011–2013. Available online: https://rieke.link/MASSIF_Architecture_document.pdf (accessed on 22 January 2020).

48. Casey, T. Threat Agent Library Helps Identify Information Security Risks. Intel. Technical Report. 2007. Available online: https://www.sbs.ox.ac.uk/cybersecurity-capacity/system/files/Intel%20-%20Threat%20Agent%20Library%20Helps%20Identify%20Information%20Security%20Risks.pdf (accessed on 22 January 2020).

49. Shanmugam, B.; Idris, N.B. Hybrid Intrusion Detection Systems (HIDS) Using Fuzzy Logic. 2011. Available online: https://www.intechopen.com/books/intrusion-detection-systems/hybrid-intrusion-detection-systems-hids-using-fuzzy-logic (accessed on 22 January 2020).

50. Dickerson, J.E.; Dickerson, J.A. Fuzzy Network Profiling for Intrusion Detection. In Proceedings of the 19th International Conference of the North American Fuzzy Information Processing Society—NAFIPS (Cat. No.00TH8500), PeachFuzz 2000, Atlanta, GA, USA, 13–15 July 2000; pp. 301–306.

51. Shanmugam, B.; Idris, N.B. Improved Intrusion Detection System Using Fuzzy Logic for Detecting Anamoly and Misuse Type of Attacks. In Proceedings of the 2009 International Conference of Soft Computing and Pattern Recognition, Malacca, Malaysia, 4–7 December 2009; pp. 212–217.

52. Chen, C.T. Extensions of the TOPSIS for group decision-making under fuzzy environment. *Fuzzy Sets Syst.* **2000**, *114*, 1–9. [CrossRef]

53. Structured Threat Information eXpression (STIX™) 1.x Archive Website. Available online: https://stixproject.github.io/ (accessed on 25 January 2020).

54. About STIX. Available online: https://stixproject.github.io/about/ (accessed on 25 January 2020).

55. Rade, R.; Deshmukh, S.; Nene, R.; Wadekar, A.S.; Unny, A. Temporal and Stochastic Modelling of Attacker Behaviour. In *Advances in Data Science 941, Proceedings of the Communications in Computer and Information Science (ICIIT 2018)*; Akoglu, L., Ferrara, E., Deivamani, M., Baeza-Yates, R., Yogesh, P., Eds.; Springer: Singapore, 2019.

56. Singapore University of Technology and Design Official Web Site. iTrust. Dataset Characteristics. Available online: https://itrust.sutd.edu.sg/itrust-labs_datasets/dataset_info/ (accessed on 25 January 2020).

57. Kotenko, I.; Chechulin, A.; Branitskiy, A. Generation of Source Data for Experiments with Network Attack Detection Software. *J. Phys. Conf. Ser.* **2017**, *820*, 012033. [CrossRef]

58. Kotenko, I.; Fedorchenko, A.; Doynikova, E.; Chechulin, A. An Ontology-based Hybrid Storage of Security Information. *Inf. Technol. Control* **2018**, *47*, 655–667.

59. Doynikova, E.; Kotenko, I. Approach for determination of cyber attack goals based on the ontology of security metrics. In *IOP Conference Series: Materials Science and Engineering (MSE) 450, Proceedings of the International Workshop "Advanced Technologies in Aerospace, Mechanical and Automation Engineering" (MIST: Aerospace-2018), Krasnoyarsk, Russia, 20 October 2018*; IOP Publishing: Bristol, UK, 2018.

 *information*

Article

Evaluation of Attackers' Skill Levels in Multi-Stage Attacks [†]

Terézia Mézešová *, Pavol Sokol and Tomáš Bajtoš

Faculty of Science, Pavol Jozef Šafárik University in Košice, 040 01 Košice, Slovakia; pavol.sokol@upjs.sk (P.S.); tomas.bajtos@student.upjs.sk (T.B.)

* Correspondence: terezia.mezesova@upjs.sk

† This paper is an extended version of our paper published in International Workshop on Systems Safety and Security—IWSSS 2019, Pitesti, Romania, 27–29 June 2019.

Received: 1 October 2020; Accepted: 17 November 2020; Published: 19 November 2020

Abstract: The rapid move to digitalization and usage of online information systems brings new and evolving threats that organizations must protect themselves from and respond to. Monitoring an organization's network for malicious activity has become a standard practice together with event and log collection from network hosts. Security operation centers deal with a growing number of alerts raised by intrusion detection systems that process the collected data and monitor networks. The alerts must be processed so that the relevant stakeholders can make informed decisions when responding to situations. Correlation of alerts into more expressive intrusion scenarios is an important tool in reducing false-positive and noisy alerts. In this paper, we propose correlation rules for identifying multi-stage attacks. Another contribution of this paper is a methodology for inferring from an alert the values needed to evaluate the attack in terms of the attacker's skill level. We present our results on the CSE-CIC-IDS2018 data set.

Keywords: alert correlation; attack evaluation; attacker skill level

---

1. Introduction

The increasing number of systems connected to the Internet presents a new set of risks for organizations as they become an interesting target not only for opportunistic attacks but targeted multi-stage attacks as well. Multi-stage attacks consist of several steps and are executed in logical follow-up steps.

Security operation centers monitor the activity within an organization's network for various threats and employ a wide range of tools to provide situational awareness to responsible asset owners. Large organizations especially grapple with a lot of legitimate network traffic, and they experience a massive number of alerts that are generated by intrusion detection systems. In such an environment, it is difficult for the analysts to filter out the noise and to discover logical relations between the alert and construct attack scenarios on a higher abstract level that the asset owners will be able to process.

One of the most common practices is to correlate events from intrusion detection systems into an attack path, a so-called multi-stage attack. These attacks are further prioritized, and the aim is to minimize the number of attacks that analysts must investigate. In threat and risk analysis, risks associated with vulnerabilities that are considered difficult to exploit are often given low priority for treatment. Therefore, analysts should be monitoring such risks and checking for any attacks targeting such vulnerabilities. They should have a comprehensive framework available so that they are able to evaluate how difficult a detected attack is to execute and treat it with the appropriate priority.

Information about attacks can be forwarded to risk management, and the appropriate countermeasures should be given a higher priority. To foster the feedback between operations and security monitoring, there should be a common understanding of the terms difficulty of an attack and difficulty of vulnerability exploitation.

Our objective in this research is to propose a set of correlation rules that will connect alerts from intrusion detection systems into more meaningful attack paths that reflect multi-stage attacks. Furthermore, a contribution of this paper is also a revised methodology first presented in [1]. It provides a common framework for inferring values needed for the evaluation of an attacker's skill level needed to exploit a vulnerability and an attacker's skill level needed or demonstrated that leads to a particular alert being raised by an intrusion detection system. It creates a relationship between what kinds of properties are evaluated for a vulnerability and how alerts can be evaluated with regard to such properties. The contribution of this paper is that we can determine the skill level needed to execute multi-stage attacks, especially the enhanced determination of how to establish the appropriate metric values on which the skill level evaluation is based.

To formalize the scope of this paper, our research objectives were:

- Define alert correlation rules to identify multi-stage attacks.
- Revise the framework for evaluating an attacker's skill level with regard to alerts.

This paper is organized into five sections. In Section 2, we discuss the related works on attacker profiles and the behaviors of attackers. Section 3 presents the rules used for alert correlation and a revised method for evaluating the skill level needed to raise alerts by an intrusion detection system. Finally, we present the results and discuss on example cases on data from a public data set in Section 4.

2. Related Works

In this section, we discuss similar works in the area of attacker skill modeling with a focus on the attackers' capabilities. In terms of quantitative analysis, Paulauskas and Garsva [2] suggested using the mean time to compromise within a normally-distributed interval as an evaluation of attacker's skill. Hu, Liu, Zhang and Zhang [3] used an absorbing Markov chain to extract various properties of attack scenarios and attackers, such as the estimated probability of reaching each attack target and the estimated occurrence number of each alert in the attack scenario. A risk estimation method that incorporates attackers' capabilities in estimating the likelihood of threats as conditions for using means and opportunities is presented by Othmane et al. [4]. They demonstrated the use of the proposed risk estimation method on video conferencing systems and connected vehicles. Additionally, they focused on evaluating whether incorporating attacker capabilities reduces the uncertainty in the experts' opinions and showed that changing attackers' capabilities changes the risks of the threats.

Rensburg et al. [5] proposed a method of generating a set of attack graphs, parameterized by attacker profiles. Vertices correspond to states of the network and an attacker, and edges correspond to actions that the attacker can take. They defined profiles as collections of capabilities and generated complete sets of profiles. This ensures analysis is not only about common types of attackers. An attacker model in networked embedded systems is presented in [6], which is used to sense, actuate and control physical processes. The authors defined an attacker framework with mapping 23 attacker profiles from related work into that framework. They defined a distance metric that allows computing overlap and discrepancies between attacker models in related work. GAMFIS [7] defines attacker types taking into account all of an attacker's attributes—motivation, skills and resources. Attacker models are further extended in [8] using types of attacker manipulation with a node, monitoring capabilities and movement strategies. Within node manipulation, they distinguish (I) attackers who are able to extract a part of the key material from the compromised node, (II) attackers who compromise the node, extracting all key material and installing their malware and (III) attackers who compromise the node and actively influence the behavior of the node. According to the attackers' capabilities, they distinguish global and local attackers.

The behavior of attackers is considered in [9]. The authors strive for a more refined attacker model introducing the attacker's view of a system. This view drives the actions of the attacker, depending on the knowledge and resources the attacker possesses. A Markov decision process models the behavior of the attacker as the method for the selection of attack steps. In [10] the attacker's skill level is classified based on the level of unauthorized access that an attacker has reached in a network. Using a multilayer fuzzy logic, attackers are divided into three categories (high, medium and low). Hassan and Guha deployed deception to characterize further the relationship between an attacker and their target [11]. They aimed to develop reliable capabilities. Their results demonstrate an association between average skill level and overall effectiveness and success of deception in unique ways.

3. Methodology

3.1. Data Set

In this paper, we worked with the data set CSE-CIC-IDS2018 published by Sharafaldin, Lashkari and Ghorbani [12]. The data set contains seven different attack scenarios: brute-force, heartbleed, botnet, denial of service, distributed denial of service, Web attacks and infiltration of the network from the inside. Infrastructure includes 50 machines belonging to the attackers, and the victim organization includes 420 machines and 30 servers. All the machines were hosted by Amazon Web Services. The data set includes the captured network traffic and system logs of each machine from several days over the course of weeks over February and March 2018.

The methodology that the authors of the data set present makes the data set suitable for various research purposes. It was used in many research papers, most recently in comparative studies on machine learning and deep learning methods for intrusion detection [13,14] or to verify the intrusion detection method for the Internet of Things environment [15].

3.2. Processing Data and Identification of Attacks

In this section, we present the methodology on processing data into security alerts and a process of attack identification. An overview of the full process is illustrated in Figure 1. Firstly, the raw data set that is available as a network capture is processed through an intrusion detection system (IDS) which generates alerts based on its available rules. Then, the alerts are input for our aggregation and correlation rules, and we are left with the identified multi-stage attacks. Finally, we evaluate each hyper-alert with the appropriate metric values and give a final evaluation of the attacker's skill level.

For data set processing, we used the intrusion detection system (IDS) Suricata (https://suricata-ids.org) with two sets of alert rules. The first rule set is one that is provided by Suricata by default. Company Proofpoint provides the second open rule set called Emerging Threats (https://rules.emergingthreats.net/). Rules from the first rule set are marked as "GPL" and rules from the second rule set are marked as "ET". The outputs from Suricata IDS were event logs, consisting of security alerts, which were further processed.

Alert processing can be divided into three stages: alert pre-processing, alert aggregation and alert correlation. At the end of the process, the results are grouped into hyper-alerts with some level of the relation between them.

At first, we selected 8 days with multiple interesting attacks, in particular Wednesdays, Thursdays and Fridays. Raw alerts from those days were normalized into the format expected by the next stages.

From the original raw alert returned by Suricata, we kept only the parameters below. Further, we filtered out noisy alerts and alerts with low severity, which we expect to be benign, and therefore, they represent false positives. It is possible that not all false positives were removed from the data, but it significantly reduced the number of noisy alerts. At this point, the alert pre-processing stage is finished.

- Date and time;
- Source and destination IP addresses;
- Source and destination network ports;
- Protocol;
- Identifier of the rule (sid);
- Rule name or message;
- Alert severity.

The second stage is alert aggregation. The network intrusion detection systems generate big amounts of log data that might or might not be of interest for security-related purposes. Aggregation and therewith the reduction of the number of the alerts are necessary steps. We split the data set into distinct days, and for each day, we ran the aggregation process separately.

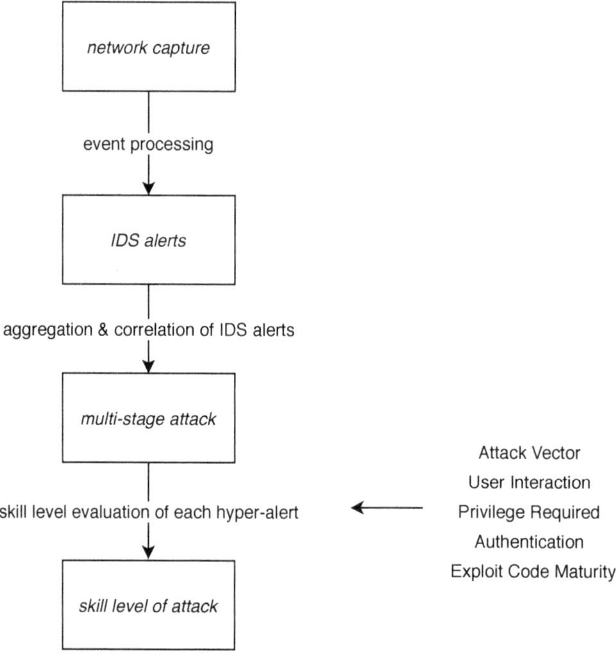

Figure 1. Overview of the processing and evaluation stages.

We use a sliding time window of size t, in which we aggregate alerts if they have the same alert sid, the same source IP address and the same destination IP address. The aggregated alerts with the time difference between the last seen alert and a new incoming alert of the same type larger than the size of the aggregation time window are excluded from the aggregation window, and a new instance of hyper-alert is created. The appropriate size of a time window *t* is between 5 and 30 min. With a bigger size of the time window, alerts which originate from different attack vectors may be aggregated together, and even more importantly, the reduction of alerts is not as significant anymore.

At the end of this stage, we have aggregated hyper-alerts that share the same features. In the structure of hyper-alerts, we left original alerts as a separate parameter, but in further processing, the whole aggregated hyper-alerts are used. The individual parameters of the alerts are united into the sets of source IP addresses, destination IP addresses and sids. The time parameter is represented in hyper-alerts as start time, which represents an occurrence of the first alert, and end time, which represents an occurrence of the last alert aggregated into the same hyper-alert. This is an example of output after the aggregation phase. In the Results section, we present only the message for clarity.

- 2018-03-02 17:50:16.251431, 114.86.88.5, 172.31.69.26, 2010939
  ET SCAN Suspicious inbound to PostgreSQL port 5432

In the third stage, the aggregated hyper-alerts were correlated according to predefined correlation rules based on an alert similarity. Each of the six predefined rules uses its own similarity function. The similarity functions take sid, source IP address and destination IP address as inputs and return a value true if the appropriate rule is fulfilled, and else return false. The rules are applied in a defined order, and two hyper-alerts are compared if they occurred in the same time window. For the size of the time window, the same applies as for the aggregation window. The rules in exact order are:

1. Each of two compared hyper-alerts has at least one identical source IP address and at least one identical destination IP address.
2. Each of two compared hyper-alerts has at least one identical sid and at least one identical destination IP address.
3. Each of two compared hyper-alerts has at least one identical sid and at least one identical source IP address.
4. Each of two compared hyper-alerts has at least one identical sid and at least one of the destination IP addresses of the first alert; and one of the source IP addresses of the second alert is the same for both.
5. Each of two compared hyper-alerts has at least one source IP address in common with the other.
6. In each of two compared hyper-alerts, at least one of the destination IP addresses of the first alert and one of the source IP addresses of the second alert are identical.

The correlation stage uses the same structure for alerts as the aggregation stage, so the alert outputs are also called the hyper-alerts. Each rule represents some similarity between alerts. For example, the fourth and the sixth rule represent a hopping of the attack from one IP address to another. The first rule represents different kinds of alerts, but between the same IP addresses. All of those rules are applied within the relatively short time window so that we can expect those individual alerts have some similarity. Therefore, we assume that the correlated hyper-alerts are sequences of alerts that belong to the same attack.

3.3. Evaluation of Attacker Skill Level of the Hyper-Alert

In this section, we present a revised methodology from [1] on evaluating the necessary skill level of an attacker from the detected alerts raised by intrusion detection systems (IDS). The methodology aims to be a corresponding partner to determining a skill level needed to exploit a vulnerability as presented in [16], so that both vulnerabilities and IDS alerts can be evaluated under the same framework. Reference [16] presents the results of an expert survey on determining an attacker's skill level needed to exploit a vulnerability or a sequence of vulnerabilities successfully. Three skill levels were defined. Parts of their definitions are also more concrete actions that attackers with said skill levels are capable of performing.

These were then in the second round of survey mapped to individual values of metrics in the common vulnerability scoring system (CVSS), as defined in [17,18]. In this paper, we are using the mappings between the CVSS metric value and skill level category, as presented in the referenced paper.

The available skill level categories TM: Confirming the changes. are from least (minimal) to most skilled (maximum): script kiddies, moderately skilled attackers and highly skilled attackers. The script kiddie category represents attackers with basic IT knowledge, and their capabilities are limited to running publicly available tools or exploits. Moderately skilled attackers can customize their toolbox to fit their attack target better, and combine various components to perform an attack. Highly skilled attackers are described as attackers with in-depth, often professional technical knowledge, and can create functional exploitative codes, which are custom-fit towards their target networks.

That paper was adapted to reflect the skill level needed to raise an IDS alert in subsequent paper [1]. Firstly, individual hyper-alerts were evaluated. In this paper, we present revised keywords and observations of an IDS alert or the alert definition and the values of the individual CVSS metrics.

Equation (1) shows that to determine the skill level of an individual hyper-alert sl_{alert}, the maximum skill level of five metrics is taken. The maximum of the evaluated skill level values for individual metrics is suitable because an attacker must demonstrate a more mature level to trigger an IDS alarm successfully.

$$sl_{alert} = max(m^{AV}, m^{UI}, m^{PR}, m^{Au}, m^{EC}) \tag{1}$$

After each hyper-alert is evaluated individually, we can evaluate the entire correlated path of hyper-alerts by choosing the maximum skill level of the present individual evaluations. We assign the lowest possible skill level to the whole path— this represents the minimum level of skill to raise the alerts detected by the IDS.

For each hyper-alert, we trace to the original alert rule according to the sid parameter, which is a unique identifier of each rule. Rules might include a reference to a vulnerability's identifier, in which case the skill level needed to trigger the alert in the network is the same as the skill level needed to exploit the originating vulnerability. IDS raise alerts based on observations of events, and they can be arbitrarily written by security analysts and often do not correspond to any particular vulnerabilities at all. A good example is reconnaissance scanning. Scanning tools such as nmap can, when configured, be used to determine whether any known vulnerabilities are present in the target network. The values of the individual metrics can be determined for a hyper-alert from the alert rules' definitions, positions of the source and destination IP addresses, how quickly the alerts are triggered and from analyst's manual determination. The values are then mapped to the skill level category, as shown in Table 1.

A hyper-alert's evaluation of script kiddies means that the least skilled attackers are able to perform the actions detected by the hyper-alert. Network hosts where such hyper-alerts are occurring in high volumes should be prioritized for patching or selected for closer monitoring. On the other hand, the occurrence of alerts evaluated with a highly skilled attacker should notify the security analysts to report a security incident and execute a response plan. The skill level metric can provide further context for the statistics reported by security teams to the stakeholders that are not familiar with the cybersecurity domain.

Table 1. Keywords or observations that distinguish skill level categories.

Keywords or Observations	Corresponding Metric Value	Skill Level
external source IP address	m^{AV}: Network	*script kiddies*
internal source & *external* destination IP address	m^{AV}: Network	*script kiddies*
internal source & destination IP address	m^{AV}: Adjacent	*moderately skilled*
small time differences in timestamps or regularity	m^{UI}: None	*script kiddies*
none - explicitly stated so references	m^{UI}: Required	*moderately skilled*
`default`	m^{PR}: None/Low	*script kiddies*
classtype *"successful-admin"* or *"Admin"* in message	m^{PR}: High	*moderately skilled*
`default`	m^{Au}: None	*script kiddies*
"Authentication Success" in message	m^{Au}: Single	*script kiddies*
more alerts with *"Authentication Success"*	m^{Au}: Multiple	*moderately skilled*
scanning or *Metasploit* in message or classtype	m^{EC}: High	*script kiddies*
malware names in message or classtype	m^{EC}: High	*script kiddies*
references in rule to ready-to-use payloads	m^{EC}: High	*script kiddies*
tagged with *SQL injection* or *Cross-site scripting*	m^{EC}: High	*script kiddies*
classtypes *rootkit* or *backdoor* (or in message)	m^{EC}: Functional/Proof of Concept	*moderately skilled*
classtype *Web-application-attack*	m^{EC}: Functional/Proof of Concept	*moderately skilled*
`default`	m^{EC}: Unproven	*highly skilled*

The attack vector metric m^{AV} defines the position of an attacker relative to the target or defended network. It can distinguish between script kiddies and moderately skilled attackers, respectively, if an attack can be performed from a remote network (e.g., Internet), or if a local connection is required.

It can be determined from hyper-alert by checking the source and destination IP addresses. If the alert is originating from an external network, which is the definition for attack vector metric value network, the skill level script kiddies is assigned. Commonly, the targeted host will respond, and this might also cause IDS to raise an alert. In the response scenario, the source IP address is internal, and the destination IP address is external. This is still within the understanding of the (m^{AV}) value network definition, and so the mapped skill level remains script kiddies. If the attack is originating from and is also targeting the internal network, an attacker has previously gained access to an internal machine in some way, and we assign the skill level moderately skilled. For the scope of this paper, we do not distinguish between an insider attacker and a truly external attacker. Outside methodologies ought to be used to recognize an insider, e.g., reference [19], and the skill level can be lowered or increased accordingly by an analyst. The attack vector metric value "local" need not be considered, as it is assigned to vulnerable components that do not have network connections [18]. Thus, no incoming or outgoing network traffic will be present for IDS to monitor.

The user interaction metric m^{UI} reflects whether a legitimate user's interaction is necessary or not to trigger the alert under evaluation. The determined value can be stored in the alert rule's metadata, so it can be adjusted if there are any changes to the alert rule conditions.

By default, the value none is taken for all alerts, thereby assuming the categorization of script kiddies. This is because scanning, automated brute-force tools or further lateral movements are executed without the need for a legitimate user's action. If there is a very small time difference in the timestamps or when we observe the alert occurring at regular intervals, this is very likely an automated tool that works on its own.

However, many initial infiltration techniques, such as phishing or malware, do require it. By keeping the skill level low by default and then searching for evidence that would support us in declaring that a legitimate user interacts with the system before IDS triggers an alert with the particular message, we stay closer to reality with the skill level evaluation. Some work has been done on automatically determining the value for the user interaction metric for vulnerabilities in [20,21]. Namely, [21] state that the word "file" was an important factor in predicting the value "required" for user interaction. IDS alert messages, however, are not as verbose as vulnerability descriptions from which the authors infer the metric value. Such alerts where user interaction is required can be identified by the security analyst manually when creating new rules, and adjusted accordingly for existing rules when the final evaluation of the skill level for a hyper-alert or correlated path seems insufficient.

Another important feature to consider is whether two-factor authentication is set up for any parts of the system monitored by the IDS. Two-factor authentication can help remedy many breaches of personal accounts and is a recommended best practice. At this point, it is necessary to look also at the human aspect of two-factor authentication and take into account the existence of the several issues (e.g., the vulnerabilities of using a third-party authentication provider) [22]. Security analysts should verify whether two-factor authentication is set up for the accounts and systems within their constituency. In such a case, they can evaluate the user interaction metric m^{UI} for the relevant rules with value required. If two-factor authentication is not enforced in the system, m^{UI} remains of the default value "none".

The privilege required metric m^{PR} aids in further evaluating the attacker's skill level. CVSS specification document defines this metric as *"the level of privilege an attacker must possess before successfully exploiting the vulnerability"* [18]. We aim to answer whether the attacker needs admin privilege to raise the observed IDS alerts. Possible research directions for integrating role-based/context-aware access control solutions can be found in [23]. That must not be confused with alerts classified as attempting to gain or successfully gaining user or administrator privilege.

The default value for this metric is considered none, meaning that the attacker can raise the alert under evaluation without any user privileges. It maps to the lowest skill level—script kiddies. The majority of scanning, malware and Web activities can be performed out of the box like this. A moderately skilled attacker is able to obtain administrative privileges to the system. We, therefore, observe alerts whose rule definitions contain the classtype "successful-admin" or keywords "admin access" in their messages [24]. An example, although not directly present in the data set we worked on, is an alert "ET MISC HP Web JetAdmin ExecuteFile admin access".

In the context of evaluating IDS alerts, the privilege required metric can be adapted as the level of privilege an attacker must possess to successfully execute the behavior leading an IDS to raise an alert. The specification document distinguishes between no authorization needed, authorization for basic tasks and authorization for significant control in the system. As such, it works with many access control models and can sufficiently reflect the deployment of role-based or context-aware access control solutions in the monitored network.

The authentication metric m^{Au} can distinguish between script kiddies and the moderately skilled category by counting the number of times any authentication takes place. By default an alert is assigned the value none, giving the lowest category—script kiddie. If in the correlated path of hyper-alerts, we have one alert containing the key phrase "authentication success" or similar, we will assign it with authentication metric value "single". It will still give us the script kiddies category. When we have two or more alerts with this message, it means that the attacker passed two or more authentication gates, and we assign the authentication metric to multiple. In such a case, the skill level raises to moderately skilled. An example alert with such keyword would be "ET POLICY VNC Authentication Successful". However, this alert was not raised in the processing of the data set we worked on for this paper.

The exploit code maturity metric m^{EC}, a temporal metric, is not part of the standard vulnerability score evaluation by a vendor when a vulnerability is publicly disclosed. It is left to the analyst to determine its value when needed and should be periodically checked. The metric has four possible values—high, functional, proof of concept, unproven—each representing the likelihood of the vulnerability being attacked, and is typically based on the current state of exploit techniques, exploit code availability or active "in-the-wild" exploitation [18].

In the context of the IDS alerts, we ought to determine whether the activity causing those alerts to be raised is a result or part of a publicly well-known exploit, implementation of a published proof-of-concept attack or evidence of an exploit for a vulnerability that was thought theoretical until now. There is common agreement that the more technical details are available about how to exploit a vulnerability, the less skilled of an attacker can successfully execute the exploit, and that increases the number of potential attackers. It is also possible that the highly skilled attackers will perform their operations in such a way that their attack steps will be obscured by legitimate traffic in the network or they will be spread throughout several days or weeks; they will obscure their IP address, thereby avoiding any correlation between the raised IDS alerts.

To evaluate the m^{EC} metric for a hyper-alert, let us assume that by default, all activity is a result of a theoretical exploit and assign the value unproven. This will yield us the highest skill level category—highly skilled attackers. Then, we are searching for evidence that will support us in deciding that the skill level category is objectively lower. The following sentences also reflect the order in which it should be executed, and once we reach the lowest skill set, we stop and do not consider further factors. References in the alert rule can link to websites describing attack steps and their detection, or explanation of the rule's logic, or a website providing the payloads which the detection rule is aiming for. They are evidence that we should evaluate at most with value proof of concept, so we are lowering the category to moderately skilled. The classtype of the IDS alerts can be useful here as well. Ready for use scanning tools, metasploit modules or malicious software do not require any additional modification, and so their exploit code maturity is high—mapping to the script kiddies category. Rootkits and backdoors might need to be compiled for the target host or even programmed from scratch by following a published proof of concept, so this maps to value functional or proof of concept and establishes the skill level as moderately skilled. Web application attacks are a broad class, and we cannot determine just one value for all of them with a reasonable degree of certainty. In general, any available tools to execute Web application attacks must be further modified to the target specifics, which is beyond the capabilities of script kiddies as defined in [16]; therefore we still are at the level of moderately skilled attackers. However, SQL or OS command injection attacks or cross-site scripting attacks are attacks will well-documented execution steps and fully autonomous tools are available, or a list of payloads to try is available. The value for m^{EC} metric for alerts with SQL injection, or cross site scripting (XSS) is high—which maps to the script kiddies level.

In any case, as it is a temporary metric, each alert with value unproven or proof of concept ought to be regularly checked if there were any exploits created and made public since the initial evaluation. The metric value should be changed accordingly.

4. Results and Discussion

In this section, we present the results of the correlation of alerts and the evaluation on four attacks from the data set.

4.1. Denial of Service Attack

The GoldenEye denial of service attack is well visible in the processed data. It was detected five times in a 30 min window from one external IP address targeting an internal machine. Data set description does not provide further details on the attack execution.

1. 5x `ET DOS Inbound GoldenEye DoS attack`

Let us present the evaluation flow of the skill level for this alert. The attack vector metric gets value network because the source IP address is external. There is a reference to the GitHub repository with the GoldenEye Python script; therefore, exploit code maturity is high. All other metrics assume the default value none. They all map to the skill level script kiddies—the maximum of all values. The mapping to the skill level is shown in Equation below. The skill level needed to raise the hyper-alert `ET DOS Inbound GoldenEye DoS attack` is script kiddies.

$$
\begin{aligned}
sl_{GoldenEye} \quad &= max(m^{AV} = Network, m^{UI} = None, m^{PR} = None, m^{Au} = None, m^{EC} = High)\\
&= max(script\_kiddies, script\_kiddies, script\_kiddies, script\_kiddies, script\_kiddies)\\
&= script\_kiddies
\end{aligned}
$$

4.2. Brute-Force Attacks

The presence of two kinds of a brute-force attack is well visible in the correlated alerts: SSH and cross-site scripting. SSH brute-force attack happened in 6 min and manifested with one occurrence of a potential SSH scan alert and one occurrence suggesting a likely brute-force attack. Both these alerts originated from the same external IP address and targeted the same internal machine.

1. `ET SCAN Potential SSH Scan`
2. `ET SCAN LibSSH Based Frequent SSH Connections Likely BruteForce Attack`

Skill level evaluation of the first alert follows the same as in the GoldenEye example. m^{EC} is high because of the keyword scan in the message. For the second alert, all the metrics apart from m^{EC} are the same as there are no indications in the alert's message or the alert's rule that a user interaction must happen, or higher privileges are required, or any authentication was successful. However, there are no references to exploit codes, so automatically it would be assigned the value unproven—meaning that only a highly skilled attacker can execute such attack. This is, in fact, not so. There are SSH brute-force tools readily available, and so we re-evaluated the metric with value high. As a result, both alerts can be raised by script kiddies.

$$
\begin{aligned}
sl_{SSHScan}, sl_{LibSSH} \quad &= max(m^{AV} = Network, m^{UI} = None, m^{PR} = None, m^{Au} = None, m^{EC} = High)\\
&= max(script\_kiddies, script\_kiddies, script\_kiddies, script\_kiddies, script\_kiddies)\\
&= script\_kiddies
\end{aligned}
$$

A cross-site scripting brute-force attack was detected within a 40-min window with 3724 alerts, each within milliseconds of each other, indicating nothing else but the execution of an automated script to check a vulnerable website for the presence of a cross-site scripting vulnerability. The data set description of the attack states they implemented script in selenium framework to attack the vulnerable application. The attack was executed from one external IP address and targeted one internal IP address.

1. 3724x `ET WEB_SERVER Script tag in URI Possible Cross Site Scripting Attempt`

This is an interesting case for an evaluation of the cross-site scripting (XSS) attempt, especially in terms of the determination of the exploit code maturity metric and user interaction metric. Let us start with the values that are straightforward to assess: The attack vector is network because of the external source IP address. It was a Web application attack, and there are no indications that a different privilege than none was needed or that any authentication had to be successful before the alert was raised. For many types of cross-site scripting, an interaction from a legitimate user in the form of clicking on a link is required. However, here we observed such small differences in the times and the number of XSS attempts was so high, that it could not be anything else but an automated brute-force attack. Therefore, user interaction was set to none. Concerning the exploit code maturity metric, the alert's rule included a reference to a list of XSS payloads, thereby providing evidence of a highly independent exploit code. The resulting skill level required to raise this alert is, therefore, script kiddies.

$$sl_{XSS} = max(m^{AV} = Network, m^{UI} = None, m^{PR} = None, m^{Au} = None, m^{EC} = High)$$
$$= max(script\_kiddies, script\_kiddies, script\_kiddies, script\_kiddies, script\_kiddies)$$
$$= script\_kiddies$$

4.3. SQL Injection Attack

Execution of an SQL injection attack was detected with altogether five different types of alerts within a 9-min window. The first alert came from an external IP address targeting an internal Web server. Responses from the Web server were sent to the external IP address as well. The nature of the alert messages and the execution in less than 10 min suggests usage of an automated tool, such as sqlmap. In the first steps, we can see an attempt to get an error in the result to verify the possible presence of a vulnerability. Then there were various SQL injection payloads tried again, resulting in an error response. In the end, we see a clear success with the last of the payloads and retrieving data from the database.

1. 2x ET WEB_SERVER SQL Errors in HTTP 200 Response (error in your SQL syntax)
2. 4x ET WEB_SERVER Possible SQL Injection Attempt UNION SELECT
3. 2x ET WEB_SERVER Possible Attempt to Get SQL Server Version in URI using SELECT VERSION
4. 2x ET WEB_SERVER Possible SQL Injection Attempt UNION SELECT
5. 2x ET WEB_SERVER SQL Errors in HTTP 200 Response (error in your SQL syntax)
6. 8x ET WEB_SERVER Possible SQL Injection Attempt UNION SELECT
7. 2x ET WEB_SERVER Possible MySQL SQLi Attempt Information Schema Access
8. 2x ET WEB_SERVER Possible SQL Injection Attempt SELECT FROM
9. 2x ET WEB_SERVER SQL Errors in HTTP 200 Response (error in your SQL syntax)
10. 2x ET WEB_SERVER Possible SQL Injection Attempt UNION SELECT
11. 2x ET WEB_SERVER Possible MySQL SQLi Attempt Information Schema Access
12. 4x ET WEB_SERVER Possible SQL Injection Attempt UNION SELECT
13. 2x ET WEB_SERVER Possible MySQL SQLi Attempt Information Schema Access
14. 2x ET WEB_SERVER Possible SQL Injection Attempt SELECT FROM
15. 2x ET WEB_SERVER ATTACKER SQLi - SELECT and Schema Columns

All of these alerts have the same attributes when it comes to determining the values for the metrics. They all are communications between an external IP address and an internal IP address, so there is no need to choose differently than the attack vector value network. There is also no evidence showing that a user interaction must take place, or that any user privilege must be acquired. The number of alerts and the short time window and regularity with which they were observed supports nothing else but the usage of an automated tool. As it is a Web application attack, we do not have enough evidence to support that any authentication must have taken place before the attack was executed. Therefore, we keep the default values for the metrics m^{UI}, m^{PR} and m^{Au}, and set m^{EC} to high to reflect that for SQL injections there are publicly available lists of payloads to try and tools that automate these checks. The final skill level evaluation needed to raise any of the observed SQL injection hyper-alerts is script kiddies.

$$sl_{SQL} = max(m^{AV} = Network, m^{UI} = None, m^{PR} = None, m^{Au} = None, m^{EC} = High)$$
$$= max(script\_kiddies, script\_kiddies, script\_kiddies, script\_kiddies, script\_kiddies)$$
$$= script\_kiddies$$

4.4. Infiltration Attacks

There were three different infiltration techniques detected in the data: usage of a file via Dropbox, usage of Eternalblue exploit and usage of a trojan via Metasploit.

The first attack was detected within a 1-min window with two occurrences of the hyper-alert `ET POLICY Dropbox.com Offsite File Backup in Use`. After the user downloaded a file via Dropbox to their machine, alerts were detecting various types of scans from the infiltrated IP address, indicating that an attacker scanned the network. This corresponds to the description of the executed attack. The infiltration and scanning were detected within an 11-min window. Correlated alerts of infiltration attack 1 are:

1. `2x ET POLICY Dropbox.com Offsite File Backup in Use`
2. `ET TROJAN Windows dir Microsoft Windows DOS prompt command exit OUTBOUND`
3. `ET SCAN Behavioral Unusual Port 135 traffic Potential Scan or Infection`
4. `ET SCAN Behavioral Unusual Port 445 traffic Potential Scan or Infection`

Alert #1 notifies that twice, a file was downloaded from a host in an external network subnet to an internal machine via the Dropbox client. Thus, the attack vector metric was set to network. There is an URL linked in references with a description of a proof of concept attack. Therefore, we can set exploit code maturity metric to proof of concept (POC). In the alert, there are no keywords or elements that would suggest choosing differently than default values for the privilege required and authentication metrics. However, by extracting information from the referenced URL, it is understood that the circumstances in which that alert was raised, were such that there had to be at-least-once authenticated access of a legitimate user via the Dropbox client. The second alert was observed 19 s after the first one. This is a wide enough gap that it probably was not done automatically, but rather that a legitimate user by their actions triggered the synchronization of their Dropbox folder. That would suggest we should set the user interaction metric to required. This and the proof of concept value for m^{EC} introduce mappings to moderately skilled attackers in the inputs for the maximum function. The overall skill level needed to raise this policy violation alert is therefore moderately skilled.

Alert #2 notified us that from an internal IP address there was outgoing traffic with characteristics of a trojan. As this is an autonomous malicious software activity, the exploit code maturity was set to high. Other metrics take their default values because there is no evidence showing otherwise in the alert's rule or references. This alert can, therefore, be triggered by script kiddies.

Alerts #3 and #4 notified us of scanning behavior from an internal IP address; only the destination port differed. This means that attack vector was set to adjacent. There was no user interaction, authentication or higher required privilege, so these metrics assumed their default values of none. Due to the presence of the keyword scan in the message, exploit code maturity was set to high. The maximum function will return an overall skill level of moderately skilled needed to raise these two alerts.

$$
\begin{aligned}
sl_{Dropbox} \quad &= max(m^{AV} = Network, m^{UI} = Requir., m^{PR} = User, m^{Au} = Single, m^{EC} = POC) \\
&= max(script\_kiddies, moderate, script\_kiddies, script\_kiddies, moderate) \\
&= moderate \\
sl_{TrojanDOS} \quad &= max(m^{AV} = Network, m^{UI} = None, m^{PR} = None, m^{Au} = None, m^{EC} = High) \\
&= max(script\_kiddies, script\_kiddies, script\_kiddies, script\_kiddies, script\_kiddies) \\
&= script\_kiddies \\
sl_{scan\#2}, sl_{scan\#3} \quad &= max(m^{AV} = Adjacent, m^{UI} = None, m^{PR} = None, m^{Au} = None, m^{EC} = High) \\
&= max(moderate, script\_kiddies, script\_kiddies, script\_kiddies, script\_kiddies) \\
&= moderate
\end{aligned}
$$

The second attack shows exploitation of the Eternalblue vulnerability in Windows. The attack was executed over 12 min from one external IP address and targeted 37 machines in the victim network altogether. Eternalblue usage is not specified in the infiltration attack description in the data set.

1.   37x `ET EXPLOIT ETERNALBLUE Exploit M2 MS17-010`

The alert detecting usage of the Eternalblue exploit contains in its rule a reference to the vulnerability CVE-2017-0143 that is targeted by this exploit code. Therefore, we will use the same values for our metrics as are defined in the CVSS score for this vulnerability. We use the CVSS string vectors from the National Vulnerability Database for this vulnerability, which are:

- CVSS:3.0/**AV:N**/AC:H/**PR:N**/**UI:N**/S:U/C:H/I:H/A:H
- AV:N/AC:M/**Au:N**/C:C/I:C/A:C

From these vectors, we use emphasized metric values. The exploit code maturity metric is missing in the base score, and we must evaluate it ourselves. Company Rapid7, which maintains the exploits in the tool Metasploit, contains in their database an entry marked with the vulnerability's CVE identification. Therefore, there exists a functional exploit code, but there is no evidence that it is ready to use as-is; rather, it must be configured with respect to the targeted network or hosts. We will set the m^{EC} metric to functional, which maps to moderately skilled attackers. As moderately skilled is a higher level than script kiddies, the maximum function will return moderately skilled.

$$
\begin{aligned}
sl_{Eternalblue} \quad &= max(m^{AV} = Network, m^{UI} = None, m^{PR} = None, m^{Au} = None, m^{EC} = Functional) \\
&= max(script\_kiddies, script\_kiddies, script\_kiddies, script\_kiddies, moderate) \\
&= moderate
\end{aligned}
$$

The third attack was detected by alerts that aggregate to just one hyper-alert. The data set's description does not provide further details on what type of exploit was executed with the Metasploit framework. We also do not see whether the infiltration through this technique was successful. The same as after the first attack, here we also see scans originating from the infiltrated internal IP address. The attack was detected within a 10-min window. The infiltration payload originated from an external IP address. The infiltrated IP address then became the source from which the attacker scanned the network.

1.   `ET TROJAN Possible Metasploit Payload Common Construct Bind_API (from server)`
2.   `ET SCAN Behavioral Unusual Port 139 traffic Potential Scan or Infection`
3.   `ET SCAN Behavioral Unusual Port 1434 traffic Potential Scan or Infection`

Alert #1 suggests a possible trojan payload coming through Metasploit tool from an external IP address. Due to the presence of that keyword in the message, the exploit code maturity was set to high. Other metrics take their default values, because there is no evidence showing otherwise in the alert's rule or references. The skill level needed to trigger this alert is therefore script kiddies. Alerts #2 and #3 are the same scanning alerts as in the first infiltration attack.

$$
\begin{aligned}
sl_{Trojan} \quad &= max(m^{AV} = Network, m^{UI} = None, m^{PR} = None, m^{Au} = None, m^{EC} = High) \\
&= max(script\_kiddies, script\_kiddies, script\_kiddies, script\_kiddies, script\_kiddies) \\
&= script\_kiddies
\end{aligned}
$$

5. Conclusions

Security operation centers monitor the activity in an organization's network and use correlation methods to create attack paths providing more comprehensive information for situational awareness. These multi-stage attacks are further evaluated and prioritized with respect to various properties.

In this paper, we presented a set of rules for correlating alerts from intrusion detection systems into more meaningful attack steps to help reduce noisy alerts that take away the focus of security analysts. Furthermore, we presented a revised methodology for evaluating the skill level of these IDS alerts. The evaluation of an attacker's skill level is a useful metric for prioritization and situational awareness. In addition to the impact-based metrics, analysts can focus on "breaking points," i.e., alerts with suddenly higher skill level evaluations. The appearance of such an alert signals that the attacker demonstrated a high-level skill set and analysts should focus their efforts on preparing appropriate countermeasures for other possible targets. In this paper, we evaluated each hyper-alert individually with such use cases in mind and did not summarize one skill level evaluation for the full correlated path.

There are open questions as to how to design the methodology for such a summation of skill level for the entire correlated path. In a real-time scenario, the path is never fully complete, and it is more useful to evaluate individual alerts as they come and reconsider potential targets of the attack based on the newly detected actions. For example, in the example scenario where first an infiltration takes place, and then the attacker uses the compromised machine to execute scans of the internal network, the scanning itself is not inherently more complicated than the infiltration; therefore, it should not increase the skill level needed for the whole attack.

The CSE-CIC-IDS2018 data set [12] was intended for verifying anomaly-based detection methods and the executed attacks were single atomic attacks. Even though one or two show characteristics of multi-stage attacks, they are only the early stages of such attacks. Therefore a more comprehensive data set is needed for more robust verification of methods. It must be designed with the intention of capturing multi-stage attacks that simulate advanced persistent threats in the network in a more realistic manner.

Author Contributions: Conceptualization, T.M., T.B. and P.S.; methodology, T.M. and T.B.; data processing, T.B. and P.S.; results analysis, T.B., P.S. and T.M; writing—original draft preparation, T.M.; writing—review and editing, T.M. and P.S.; supervision, P.S.; All authors have read and agreed to the published version of the manuscript.

Funding: This research is funded by projects VVGS-PF-2020-1436 and VVGS-PF-2020-1427, and Slovak APVV project under contract number APVV-17-0561.

References

1. Mézešová, T.; Sokol, P.; Bajtoš, T. Evaluation of Attacker Skill Level for Multi-stage Attacks. In Proceedings of the 2019 11th International Conference on Electronics, Computers and Artificial Intelligence (ECAI), Pitesti, Romania, 27–29 June 2019; pp. 1–6.
2. Paulauskas, N.; Garsva, E. Attacker skill level distribution estimation in the system mean time-to- compromise. In Proceedings of the 2008 1st International Conference on Information Technology, Gdansk, Poland, 18–21 May 2008; pp. 1–4.
3. Hu, H.; Liu, Y.; Zhang, H.; Zhang, Y. Security metric methods for network multistep attacks using AMC and big data correlation analysis. *Secur. Commun. Netw.* **2018**, *2018*. [CrossRef]
4. ben Othmane, L.; Ranchal, R.; Fernando, R.; Bhargava, B.; Bodden, E. Incorporating attacker capabilities in risk estimation and mitigation. *Comput. Secur.* **2015**, *51*, 41–61. [CrossRef]
5. van Rensburg, A.J.; Nurse, J.R.; Goldsmith, M. Attacker-parametrised attack graphs. In Proceedings of the Tenth International Conference on Emerging Security Information, Systems and Technologies, Nice, France, 24–28 July 2016.
6. Rocchetto, M.; Tippenhauer, N.O. On attacker models and profiles for cyber-physical systems. In Proceedings of the European Symposium on Research in Computer Security, Heraklion, Greece, 26–30 September 2016; Springer: Berlin/Heidelberg, Germany, 2016; pp. 427–449.
7. Fraunholz, D.; Anton, S.D.; Schotten, H.D. Introducing gamfis: A generic attacker model for information security. In Proceedings of the 2017 25th International Conference on Software, Telecommunications and Computer Networks (SoftCOM), Split, Croatia, 21–23 September 2017; pp. 1–6.
8. Ošt'ádal, R.; Švenda, P.; Matyáš, V. Reconsidering attacker models in Ad-Hoc networks. In Proceedings of the Cambridge International Workshop on Security Protocols, Brno, Czech Republic, 7–8 April 2016; Springer: Berlin/Heidelberg, Germany, 2016; pp. 219–227.
9. Krautsevich, L.; Martinelli, F.; Yautsiukhin, A. Towards modelling adaptive attacker's behaviour. In Proceedings of the International Symposium on Foundations and Practice of Security, Montreal, QC, Canada, 25–26 October 2012; Springer: Berlin/Heidelberg, Germany, 2012; pp. 357–364.
10. Mohammadian, M. Intelligent security and risk analysis in network systems. In Proceedings of the 2017 International Conference on Infocom Technologies and Unmanned Systems (Trends and Future Directions) (ICTUS), Dubai, UAE, 18–20 December 2017; pp. 826–830.
11. Hassan, S.; Guha, R. A probabilistic study on the relationship of deceptions and attacker skills. In Proceedings of the 2017 IEEE 15th International Conference on Dependable, Autonomic and Secure Computing, 15th International Conference on Pervasive Intelligence and Computing, 3rd International Conference on Big Data Intelligence and Computing and Cyber Science and Technology Congress (DASC/PiCom/DataCom/CyberSciTech), Orlando, FL, USA, 6–10 November 2017; pp. 693–698.
12. Sharafaldin, I.; Lashkari, A.H.; Ghorbani, A.A. Toward generating a new intrusion detection dataset and intrusion traffic characterization. In Proceedings of the 4th International Conference on Information Systems Security and Privacy (ICISSP 2018), Funchal, Portugal, 22–24 January 2020; pp. 108–116.
13. Gamage, S.; Samarabandu, J. Deep learning methods in network intrusion detection: A survey and an objective comparison. *J. Netw. Comput. Appl.* **2020**, *169*, 102767. [CrossRef]
14. Ferrag, M.A.; Maglaras, L.; Moschoyiannis, S.; Janicke, H. Deep learning for cyber security intrusion detection: Approaches, datasets, and comparative study. *J. Inf. Secur. Appl.* **2020**, *50*, 102419. [CrossRef]
15. de Souza, C.A.; Westphall, C.B.; Machado, R.B.; Sobral, J.B.M.; dos Santos Vieira, G. Hybrid approach to intrusion detection in fog-based IoT environments. *Comput. Netw.* **2020**, *180*, 107417. [CrossRef]
16. Mézešová, T.; Bahsi, H. Expert Knowledge Elicitation for Skill Level Categorization of Attack Paths. In Proceedings of the 2019 International Conference on Cyber Security and Protection of Digital Services (Cyber Security), Oxford, UK, 3–4 June 2019; pp. 1–8.

17. Mell, P.; Scarfone, K.; Sasha, R. A Complete Guide to the Common Vulnerability Scoring System Version 2.0. Published by FIRST-Forum of Incident Response and Security Teams, 2007. Available online: https://tsapps.nist.gov/publication/get_pdf.cfm?pub_id=51198 (accessed on 19 November 2020).
18. CVSS Special Interest Group. Common Vulnerability Scoring System v3.1: Specification Document; 2019. Available online: https://www.first.org/cvss/v3-1/cvss-v31-specification_r1.pdf (accessed on 19 November 2020).
19. Andel, T.R.; Yasinsac, A. Adaptive threat modeling for secure ad hoc routing protocols. *Electron. Notes Theor. Comput. Sci.* **2008**, *197*, 3–14. [CrossRef]
20. Allodi, L.; Banescu, S.; Femmer, H.; Beckers, K. Identifying relevant information cues for vulnerability assessment using CVSS. In Proceedings of the Eighth ACM Conference on Data and Application Security and Privacy, Tempe, AZ, USA, 19–21 March 2018; pp. 119–126.
21. Elbaz, C.; Rilling, L.; Morin, C. Fighting N-day vulnerabilities with automated CVSS vector prediction at disclosure. In Proceedings of the 15th International Conference on Availability, Reliability and Security, Dublin, Ireland, 25–28 August 2020; pp. 1–10.
22. Watters, P.; Scolyer-Gray, P.; Kayes, A.; Chowdhury, M.J.M. This would work perfectly if it weren't for all the humans: Two factor authentication in late modern societies. *First Monday* **2019**, *24*.. [CrossRef]
23. Kayes, A.; Rahayu, W.; Watters, P.; Alazab, M.; Dillon, T.; Chang, E. Achieving security scalability and flexibility using Fog-Based Context-Aware Access Control. *Future Gener. Comput. Syst.* **2020**, *107*, 307–323. [CrossRef]
24. ET Labs. *Emerging Threats Rules*; ET Labs: Austin, TX, USA, 2020.

 *information*

Article

Smali$^+$: An Operational Semantics for Low-Level Code Generated from Reverse Engineering Android Applications

Marwa Ziadia [1,*], Jaouhar Fattahi [1], Mohamed Mejri [1] and Emil Pricop [2]

[1]   Department of Computer Science and Software Engineering, Laval University, Pavillon Adrien-Pouliot 1065, avenue de la Médecine, Quebec City, QC G1V 0A6, Canada; marwa.ziadia.1@ulaval.ca (M.Z.); jaouhar.fattahi.1@ulaval.ca (J.F.); mohamed.mejri@ift.ulaval.ca (M.M.)

[2]   Automatic Control, Computers and Electronics Department. Petroleum-Gas University of Ploiesti, 100680 Ploiesti, Romania; emil.pricop@upg-ploiesti.ro

*   Correspondence: marwa.ziadia.1@ulaval.ca

Received: 31 January 2020; Accepted: 26 February 2020; Published: 27 February 2020

Abstract: Today, Android accounts for more than 80% of the global market share. Such a high rate makes Android applications an important topic that raises serious questions about its security, privacy, misbehavior and correctness. Application code analysis is obviously the most appropriate and natural means to address these issues. However, no analysis could be led with confidence in the absence of a solid formal foundation. In this paper, we propose a full-fledged formal approach to build the operational semantics of a given Android application by reverse-engineering its assembler-type code, called Smali. We call the new formal language Smali$^+$. Its semantics consist of two parts. The first one models a single-threaded program, in which a set of main instructions is presented. The second one presents the semantics of a multi-threaded program which is an important feature in Android that has been glossed over in the-state-of-the-art works. All multi-threading essentials such as scheduling, threads communication and synchronization are considered in these semantics. The resulting semantics, forming Smali$^+$, are intended to provide a formal basis for developing security enforcement, analysis and misbehaving detection techniques for Android applications.

Keywords: Android applications; multi-threading; operational semantics; reverse engineering; Smali$^+$

---

1. Introduction

A few years ago, mobile phones were used to make calls or send messages. Today, they surpass computers as the most commonly used digital device. They manage our agenda, emails, credit cards, itineraries and business documents. Android is the most popular operating system for mobiles and embedded devices, having the largest application market and 85% of all smartphones sold in 2019 were equipped with an Android OS [1]. Android is an open nature platform, which means that applications could be downloaded from sources other than the official Google play store. This is an important feature that has contributed to its unquestionable success, given the breadth of the available application that draws people to the platform, making it an ideal target for malicious application downloads.

Indeed, users are increasingly exposed to attacks targeting the Android environment via malicious applications. They thus endanger privacy information, by disclosing sensitive data (FakeNetflix malware [2]) or collecting sensitive banking information, especially with the increasing use of banking applications (Anubis trojan [3]). Furthermore, the installation of apparently legitimate malicious applications can lead to: clandestine eavesdropping on telephone conversations; tracking GPS position; exploiting pay services to cause financial losses to the user for the benefit of the attacker by calling or sending SMS messages to premium-rate numbers without the user's knowledge (SMS Trojan such as FakePlayer, AsiaHitGroup and GGTracker [4–6].

To deal with this, automated tools for analyzing, verifying and enforcing the security of Android applications are highly needed [7–10]. Nevertheless, they must be based on a formal specification of the target platform to give solid results. In this paper, we propose formal operational semantics for a subset of the low-level Android code, which we consider particularly relevant for modeling Android applications and which we call Smali$^+$. It includes the main bytecode instructions of Dalvik, and a few important API methods related to Java concurrency. Smali$^+$ is ultimately written from Smali with some essential native methods that were replaced with macro-instructions for simplification. Smali$^+$ is intended to serve as a basis for further analysis of Android applications and security implementation techniques. Android applications are mainly written in Java. The Java source code is first compiled into a Java Virtual Machine (JVM) bytecode using a standard Java compiler called *Javac*. Following this, the Java source files are converted into class files that store Java bytecode. The Java bytecode is then translated to an optimized bytecode called *Dalvik* through a tool called *dx*. At this stage, all the class files are converted and consolidated into a single DEX file called Dalvik EXecutable or simply a DEX to save memory. An Android Package Kit (APK) is essentially a zip of the DEX file accompanied by a Androidmanifest.xml file, a set of resources and potentially shared libraries. Figure 1 illustrates these steps.

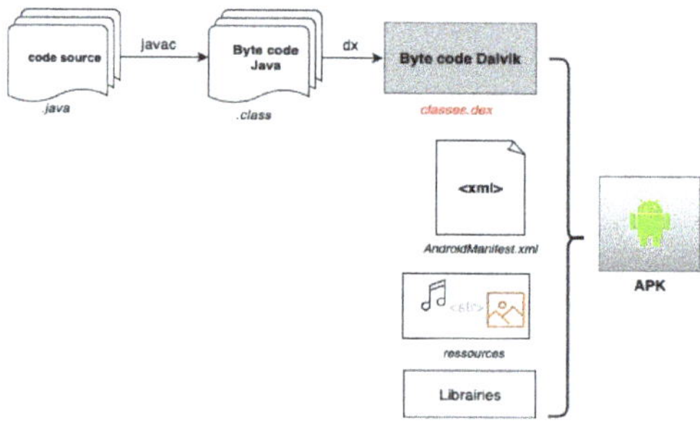

Figure 1. Compilation steps of an Android application.

In this work, we focus on the DEX format file, which contains the Dalvik binary code used even by the successor of Dalvik (since Android 5.0) called Android Runtime (ART).

Formalizing a low-level code, rather than high-level Java source or intermediate level Java bytecode, is our choice for many reasons. Firstly, Dalvik byte code is always available and it is easily obtainable from any Android application. Secondly, Dalvik bytecode is the common executable format for all Android applications and therefore the code is much closer to the code really executed. Even though decompilation from Dalvik back to Java or to Java bytecode is possible using reverse engineering tools (such as dex2jar and ded), there is no guarantee to recover the original source code since there is not a 100% robust and correct Dalvik-to-Java reverse translation tool [11]. However, even though that it is possible to retrieve source code or Java bytecode from Dalvik, editing or improving code at this level requires the user to reconvert it back to Dalvik and running the application afterward will often fail [9]. Focusing directly on Smali will avoid such problems. Hence, binary code obtained at this level, in DEX file, is illegible and requires conversion into a more understandable format prior to being analyzed, improved or edited. Reverse engineering in software makes it possible to convert a machine-readable binary file into a human-readable file, which is the case with DEX files.

Apktool [12] is a reverse engineering tool that simplifies the entire process of assembling and disassembling Android applications. It includes *"Smali"* and *"bakSmali"*, which are equivalent to *"assembler"* and *"disassembler"*, respectively allowing the passage from and to the DEX format. Apktool allows the user to disassemble applications to nearly original form. It uses *BakSmali* to produce, from an APK, a human-readable format akin to assembly languages called Smali (Smali is both the name of a mnemonic language for the Dalvik bytecode and its assembler version.). This code is nothing but a translation of the machine code generated by the DVM. In other words, it is a readable representation of Dalvik bytecode in an assembly-like code, with mnemonic instructions. *BakSmali* creates a Smali file for each class in the application preserving the original signature. The structure of such a file is presented in Figure 2. In addition to the code contained in the classes.dex file, Apktool generates the application decoded resources, as well as the *AndroidManifest.xml* file (in a readable version. These reverse engineering analysis techniques are still effective with the newly introduced ART environment [13].

```
1      .class modifiers Lsome/package/Someclass;
2      .super Lsome/package/Someclass;
3      .implements Lsome/package/Someinterface;
4      .source "someclass.Java"
5
6      .field modifiers fieldname : type;
7
8      .method modifiers methodname (type,...)type
9          .locals ...
10         instruction ...
11         instruction ...
12         instruction ...
13         ...
14     .end method
15     ...
```

Figure 2. Structure of a Smali file.

In this paper, we put forward formal semantics for Smali. Smali is an assembly-like language that runs on Android's DVM. It is obtained by 'bakSmaling' the Dalvik executable file (.dex). A syntax and semantics have been adopted to specify this low-level code. The resulting formal language is a sub-language of Smali and a simpler language, called Smali^{+}. A set of the most used Dalvik instructions have been generalized into 12 semantically different instructions (see [11] for generalization process), compared to more than 200 Dalvik original instructions in Smali. In addition to this set, our semantics includes instructions related to multi-threading. We plan to use Smali^{+} in the near future to specify security properties for Android applications and this in order to protect the user from security threats that target the Android environment through downloaded applications.

The paper is organized as follows. In Section 2, we present some related work with similar ideas of bytecode formalization and we discuss their advantages as well as their drawbacks and limitations. In Section 3, we give some essential preliminaries related to Smali (registers, some adopted notations, types, etc.). In Section 4, we present the operational syntax and semantics of Smali^{+} for a single-threaded application. In Section 5, we present the operational syntax and semantics of Smali^{+} for a multi-threaded application. In Section 8, we conclude and we introduce the future avenues of our research.

2. Related Work

Mostly, the studies based on formal semantics of Android target a single well-defined goal. This can be an analysis for certification, a detection of potential vulnerabilities or malicious behavior of an application, or a verification of any aspect. It can also be a means to reveal security breaches of Android applications [14]. We will see in the studies we are presenting hereafter that formalization elements substantially differ from one objective to another. This being said, it is practically impossible to evaluate the efficiency of analyzes that are not based on the formal specification of the targeted platform.

In [15], Payet et al. define operational semantics for a subset of Dalvik opcodes that present registers manipulation, arithmetic operations, object creation, access and method calls as well as

Android activities. Semantics rules were relatively complex. An Android program was modeled as a graph of blocks where each block has one or more instructions among the selected instructions. Blocks are linked in a way that they express control flow passing from one block to another. They require that invoke and return instructions only occur at the beginning and the end of a block, respectively. Blocks of semantics integrate instruction semantics for those that are different from a call or a return. Call instruction semantics allow passing from the caller method block to the callee method block. Activity semantics depend on the activity state, method callback, activity life cycle and external events. These semantics are defined to be the basis of static analyses that take into account the life-cycle of the activities. Despite the importance of thread-activity connection in Android semantics, threading was detached from activities semantics and concurrency was ignored in this work.

In [16,17], the authors propose a formal operational semantics for the Dalvik bytecode. The formalization was accompanied by a control flow analysis to detect potential malicious actions. Although the results highlight threading as the most often used language features with a (90.18%), this feature was omitted in both analyses and semantics to focus, instead, on reflection, exceptions and dynamic dispatch with 73.00% and 19.53%, respectively, which we find somewhat awkward. This motivates us to pay a special attention to the mutli-threading aspect modeling for Android.

In [11], the authors present SymDroid as a Dalvik bytecode interpreter for eventual security vulnerabilities detection. It is a symbolic execution for a simplified intermediate language of a fraction of Dalvik opcodes, named μ-*Dalvik*. SymDroid receives the Dalvik bytecode (the .dex file) as input. The opcode is first translated to μ-*Dalvik*, which one is based on 16 instructions considered as the most relevant ones to perform code analysis. Then, it is processed by a symbolic execution core using the SMT solver to generate traces as an intermediate result. Finally, the post-analyzer inspects the output traces and determines the final result. Entry points and all possible events affecting the application's behavior were developed according to a client-oriented specification (it is up to the user to model it) to drive the application under test as desired. Although this work's models, in addition to modeling bytecode instructions, the system libraries including Bundle and Intent, Android components life cycle, services and views; it ignores the system's concurrent nature, either in the selected bytecode instructions or at the program symbolic execution level, which is considered as being sequential.

In the same vein, Julia presented in [18] is a static analyzer for Java bytecode based on abstract interpretation. It was extended in [19] and adapted to analyze Dalvik bytecode and handle specific features of Android such as event-driven nature, potentially concurrent entry points and dynamic inflation of graphical views. It applies several static analyses for Android applications' classcast, nullness, dead code and termination analysis, but does not track information flow. Multi-threaded applications were not included in this work and event handlers are executed by a single thread.

Gunadi et al. [20,21] propose an operational semantics of DEX bytecode for certifying non-interference properties through type system. This study includes a translating process from Java bytecode semantics developed in [22] to Dalvik bytecode, concluding that if the first type system guarantees non-interference then its translation into Dalvik bytecode is also typable. Therefore existing bytecode verifiers for Java could certify non-interference properties of Dalvik bytecode.

Multi-threading programming semantics in applications have lately drawn increasing attention. Some combine it with event handling [23–25], others consider the main API methods relating to it [26]. In [24], Kanade proposes a semantic of a combined concurrency model of threads and events. All the focus in this work goes to the event-driven nature of Android and its relationship with the application's threads. As a consequence, all other states that semantics could reach, such as those resulting from basic instruction execution (method call, jump, return instruction, etc.), have not been treated. The semantics proposed in [26] were the closest to ours. They cover the main important Dalvik instructions and handle multi-threading. This paper could be seen an extension of [27], with the obviously major change of the semantics needed for the concurrent setting and exception handling. However, thread scheduling was not discussed and thread spawning is left to the virtual machine to execute in an unpredictable point in time.

In the same stream of thought, in [28], Chaudhuri presents a formal security study on Android using operational semantics and a system of types for specific Android constructs. However, semantics ignore all Java constructs that may appear in Android applications (no class and method modeling), to focus instead on Android components, intents and all Android-specific features related to it (binding a service, sending an intent, etc.). This can be seen as a unified formal understanding of security for users and developers of Android applications to deal with their security concerns.

Some works have focused on other issues of Android such as multi-tasking. For instance, ASM presented in [29] is a formal model that formalizes all Android elements related to multi-tasking, such as activities, back stacks and tasks. An Android application is somewhat seen as a collection of activities with different types that interact with the user through a back stack. ASM has recently been extended in [30] to capture all the core elements of the multi-tasking mechanism used in inter-component communication.

Over time, formalization has included the permissions system as well [31–33]. For example, Bagheri et al. propose in [31] a formal specification for Android application's permission system through an ad-hoc specification language called Alloy. It aims to formally specify the behavior of Android applications, in particular, the mutual interaction between applications based on permissions and security consequences caused by it or what authors call inter-app permission leakage vulnerabilities. Almost all Android elements related to inter-app permissions were taken into account in the formalism. Every application is modeled as a set of components, permissions, intent filters and vulnerable paths. Similarly, in [33], a formal model of the Android's permissions is specified in the theorem prover Coq syntax.

Acteve++ [34] is an automated testing tool for Android Apps. It is based on Acteve [35] but is improved to support input events and broadcast events in order to achieve higher coverage. Authors use a non-standard operational semantics that describes the concolic execution of the program. Semantics describe program execution in response to a sequence of events generated automatically from an external environment. All other features and instructions that Android handles were neglected to focus instead on the event-driven paradigm, which we found not expressive enough to model an Android application. Our operational semantics consider, besides the concurrent feature, a variety of instructions that models methods invocations, object creation and the whole tree structure of an application (class, method and fields).

In [36], the authors focused on the low-level interactions with the operating system, by recording the system calls (syscalls) invoked. To benefit from two levels, the analysis uses generic low-level syscall traces to reconstruct the high-level semantics. While syscalls analysis offers more security guarantees, it, in our opinion, complicates the task more. Especially, this information is extracted from internal interfaces between the Android libraries and the kernel, which may change in the next versions of Android without notice. In our work, we propose a rich semantics that covers all API calls at a high level and we consider that it is sufficient to enforce security policies later.

Some studies like those conducted by Stowway [8] and Comdroid [37] for flow analysis directly analyze the disassembled DEX file for a given application to identify potential component and/or communication vulnerabilities. Despite the promising results of both tools in analyzing Dalvik bytecode and Android's API, proving its soundness and evaluating its efficiency or deficiency is practically impossible in the absence of formal specification and proof.

Concurrent programming concepts and techniques are widely used in Android in order to manage different tasks and threads. Our formalism Smali$^+$ consider this important feature that was neglected before given its complexity. Overall, none of the aforementioned studies, including those considering multi-threading, offer complete semantics covering all the states that a thread can reach nor representing all multi-threading essentials. Most of the studies formalizing Dalvik byte code and handling multi-threading include only the two Dalvik instructions related to monitor use, *monitor-enter* and *monitor-exit*, since Dalvik opcodes encompass only these two instructions with regard to threading. However, a semantic for an Android program should not be limited to these instructions and must also consider instructions related to threads communication, signaling and scheduling. In this paper, we fill this gap by proposing semantics that incorporate, in addition to Dalvik instructions, a wide range of API methods covering multi-threading essentials formulated in macro instructions for the sake of simplicity. In comparison with all test-based approaches, Smali$^+$ is based on formal methods with their foundation in mathematical logic, allowing us to achieve rigorous and unambiguous reasoning in the system specification and proofs, ensuring the system proprieties, while test-based approaches can only ensure that systems satisfy the requirements for test cases. In sum, the proposed formal language is expressive enough to enforce security proprieties and to detect security critical APIs (i.e., those related to sensitive data access such as camera, SMS, telephony and contact list). Its syntax includes the class fully qualified name for each invoked method facilitating to localize such APIs.

3. Preliminaries

In this section, we present the most essential information for Smali. First, we present the DVM architecture and how it affects Smali syntax. Then, we present method invocation and how it affects Smali registers. Finally, we present Smali special notations for types.

3.1. Registers

Being optimized to run on devices on which resources and processor speed are scarce and the DVM architecture is register-based. Local variables are assigned to any of the 2^{12} available registers. A register is used to hold any data value, except for *double* and *long* values where each one requires two registers (64 bits). The Dalvik opcodes operate on the register's content instead of operating directly on values and accesses elements on a program stack such as stack-based virtual machines. Hence, registers allow the DVM to keep track of program evolution while it executes bytecode [38]. Each method in Smali has its own set of registers for each method's arguments, local variables and a special register for its return value. We will see later that most of the instructions include source and destination registers. Smali language denotes each set of registers differently, which allows us to visually distinguish between the method's local and argument registers.

The alternate *.locals* directive specifies the number of local registers used by the method (non-parameter registers) which is statically known. Local registers in Smali are denoted with $v_0, v_1, v_2, ..., v_n$, where v_0 is the first local register, v_1 the second and so on until the last register. This includes a special register for a method return value that allows passing return values from the callee back to the caller, which one is denoted by *ret*.

$$LocalRegisters = \mathbb{N} \cup \{ret\}$$

Parameter registers in Smali are denoted by $p_0, p_1, p_2, ..., p_n$. The first parameter for non-static methods is always the object that the method is being invoked on, in this case p_0 holds the object reference and p_1 the second parameter register. For a static method invocation p_1 is the first parameter register. For more details, please see the Method invocation subsection.

The *.registers* directive specifies the total number of registers in the method. This includes the registers needed to hold the method parameters, which are stored in the last registers in the method.

$$Registers = LocalRegisters \cup ParameterRegisters$$

3.2. Method Invocation

The DVM conforms to the ARM's calling convention which is used for low-level code where parameters, return values, return addresses and scope links are placed in registers. It dictates how these elements are shared between the caller and the callee. In fact, these two share a part of their register array so that the caller passes arguments to the callee by setting its parameter registers in the right order. As for class methods, a lookup procedure starts by searching in the list of all static methods that belong to the named class, where classes have distinct names and locating the invoked method through its signature (i.e., name, argument types and number, and return type). Then, its parameter registers array is set according to ARM's calling convention, so that the first argument leads to the first parameter register p_1 and so on until the last argument which identifies the last register for arguments (n arguments lead to n parameter registers).

In the dynamic invocation case, the class of the object whose method is being called (or recipient object's class) is statically unknown, so it is first retrieved from the heap through its reference (see the semantics section for more details). Then, a lookup procedure searches among the class method list upwards to its super-class chain, for a method matching the given method signature. Registers comprise an additional register for the object reference called p_0 in Smali code. Hence, the actual number of parameter registers is $p + 1$.

Local register contents are initially undefined (registers are untyped in Dalvik), however, its number is statically known.

3.3. Types in Smali

Smali code has two major classes of types, primitive types and reference types.

A primitive notation in Smali is particular where a single letter specifies each type, for example V is used for a void type.

Reference types are objects (i.e., class type) and arrays. A class type takes the form *Lpackagename/ClassName;* where the leading L indicates that it is a class type, *packagename* is the package name path where class *ClassName* belongs to, whereas *ClassName* refers to the class name. For example, a thread object in Smali has the following type: *LJava/lang/Thread;* which is equivalent to *Java.lang.Thread* in Java. Arrays take the form [*Type* (*Type* which could obviously be a primitive or a reference). Arrays with multiple dimensions are presented by corresponding number of "[" characters. For example, a two-dimension arrays of int(s) is presented as follow [[I which is equivalent to *int*[][] in Java. Table 1 summarizes different types in Smali.

Table 1. Types in Smali.

Primitive Types	
B	byte
C	char
F	float
I	int
J	long
S	short
V	void
Z	boolean
Reference Types	
Lpackagename/Classname;	Object
[... Object or Primitives	Array

4. Operational Semantics for a Single-Threaded Application

4.1. Notations

Throughout the paper, we use the following notations:

- $A :: B :: C$ to designate a stack, where A is the top-most value of the stack, B is the underlying element and C is the remaining portion of the stack. An empty stack is presented by ϵ.
- $\perp$ to denote any undefined value.
- $dom(f)$ is domain of a function f. The notation $dom[f \mapsto x]$ expresses the domain dom where the value of a function f is updated to x.
- $f[x \mapsto y]$ expresses the function f where value x maps to y so $f(x) = y$.

4.2. Syntax

Table 2 provides basic syntactic categories as well as the selected instructions syntax.

A package of a disassembled DEX bytecode format is specified by a name *pck* and sequences of classes. In our formal model, we consider that a package consists only of classes that correspond to *.Smali* files (Androidmanifest file and the rest of XML files are not considered in our formalization).

Table 2. Smali$^+$: sequential execution.

(Package)	*Pckg*	::= **.package** *pck* {*Cl**}	
(Class definition)	*Cl*	::= **.class** (*Acc-flg**) *Cfn* **.super** *Sc* **.implements** *Intf** {*Fld**, *Mtd**}	
(Super class)	*Sc*	::= *Cfn* | ⊤	
(Interface definition)	*Intf*	::= **.interface** (*Acc-flg**) *Inf* **.super** *Sinf** {*CstFld**, *MtdSign**}	
(Super interface)	*Sinf*	::= *Inf*	
(Field definition)	*Fld*	::= **.field** (*Acc-flg*)* *f* : *τ*	
(Constant Field definition)	*CstFld*	::= **.field public final static** *f* : *τ*	
(Method definition)	*Mtd*	::= **.method** (*Acc-flg*)* *MtdSign* **.locals** *loc* {*LabelInst**}	
(Method signature)	*MtdSign*	::= *m* (*τ*$_1$, ... *τ*$_n$) *retτ*	
(Access flags)	*Acc-flg*	::= **public** | **private** | **protected** | **final** | ...	
(Labeled Instruction)	*LabelInst*	::= *i Inst*	
(Label)	*i*	::= **int**	
(Instructions)	*Inst*	::= **goto** *i*	(unconditional jump)
		| **move** *Des Src*	(move from source to destination)
		| **binop**$_⊕$ *v v*$_1$ *v*$_2$	(binary operation)
		| **unop**$_⊙$ *v v*$_1$	(unary operation)
		| **if**$_⊗$ *v*$_1$ *v*$_2$ *i*	(conditional jump)
		| **new-instance** *v Cfn*	(object creation)
		| **invoke-static** *Cfn MtdSig v**	(static method invocation)
		| **invoke-instance** *v*$_{ref}$ *MtdSig v**	(instance method invocation)
		| **return** *v*	(retrun from non-void method)
		| **return-void**	(retrun from a void method)
(Destination register)	*Des*	::= *v*	(register name)
		| *v*$_{ref}$.*f*	(instance field)
		| *Cfn.f*	(static field)
(Source register)	*Src*	::= *Des* | *Cst*	(des or constant)
(Operators)	⊕	::= **+** | **-** |...	(binary operator)
	⊙	::= **¬** |**++**|...	(unary operator)
	⊗	::= **<**|**>**|...	(comparison operator)
(Program counter)	*i*	::= **int**	
(Num. of loc. registers)	*loc*	::= **int**	
(Local registers name)	*v*	::= **string**	
(Parameter registers name)	*p*	::= **string**	
(Constant)	*Cst*	::= *Single*	(constant)
(Type)	*τ*	::= *Prim* | *Ref*	
	Prim	::= *Single* | *Double*	
	Ref	::= *Cfn* | *ArrayType*	
	ArrayType	::= *ArrayTSingle* | *ArrayTDouble*	
	ArrayTSingle	::= **array** (*Single* | *Ref*)	
	ArrayTDouble	::= **array** *Double*	
	Single	::= **boolean** | **char** | **byte** | **short** | **int** | **float**	
	Double	::= **long** | **double**	
(ReturnType)	*retτ*	::= *τ* | **V**	
(Names)	*Cfn*	::= **L***packagename*|*c*	(class full name)
	Inf	::= **L***packagename*|*itf*	(interface full name)
	pck, c, itf, f, m	::= **String**	(package, class, interface, field and method names)

A class *Cl* definition includes its access flags *Acc-flg*, which is a keyword defining the class visibility, a fully qualified class name *Cfn* that indicates the class package path name followed by the class name *c* (we assume an unlimited supply of distinct names). This includes also its direct super-class fully qualified name (a single inheritance). $\top$ is applied to classes without super-classes such as the Object class and the Thread class, and finally a set of implemented interfaces *Intf*, fields *Fld* and methods *Mtd*.

An interface is specified by its fully qualified name *Inf*, access flags *Acc-flg*, a set of super-interfaces *Sinf*, its abstract methods (which consist of their method signatures) and constant fields. A field definition comprised its name *f*, its access flags and a type τ (which could be a primitive for static fields or a class type for instance fields). A method definition includes a set of access flags that determines its scope, the method signature, the number of local registers it operates on denoted by *loc* and a sequence of labeled instructions *Inst* that present the method body. A method signature consists of the method name *m*, argument(s) type τ and a return type *retτ* which might be a void, primitive or a class type. In Smali$^+$, we consider a subset of Dalvik instructions being selected based on results of a study of 1700 Android applications, carried out to determine what instructions and language features are most often used in typical applications [16,17]. In fact, Dalvik bytecode comprises 218 instructions [39]. We bring some modification to the selected instructions that does not affect the expressive power of Dalvik language. In contrast, it simplifies the representation of our semantics. For example, in Dalvik we find 13 variants of the *move* instruction that are semantically similar, we model this group of instructions by only one *move* instruction.

In our formal model, we consider instructions expressing the unconditional and conditional jump with, respectively, *goto* and *if$_\otimes$* instructions. A *move* instruction to move values from source *Src* to destination *Des*. A destination may be a register name *v*, an instance field $v_{ref}.f$ or a static field *Cfn.f*, whereas a source *Src* may be any of these elements beside constants *cst*. We consider also instructions expressing the creation of a new object of a class *Cfn*, a return from a void and non-void method with *new-instance*, *return-void* and *return* instructions, respectively. Method invocation refers to the method name, argument types and number, return type and registers. For methods class that are dynamically dispatched, it includes in addition to that a register holding the recipient object reference.

4.3. Semantics

Table 3 defines the domains used by our operational semantics. In fact, each application has at least one thread that defines the code path of execution and all of the code will be processed along the same code path if there is no other created thread. Hereafter, we suppose a single-threaded execution, a simple programming model with deterministic execution order, which means that an instruction has to wait for all preceding instructions to finish prior to being processed. We model such execution with a local configuration denoted by σ. It models the full state of a single-threaded program. It includes a call stack C_s, a heap *H* and a static heap *S*. A call stack allows keeping track of all information concerning methods invoked in the program. It is initially empty and presented as a sequence of method frames. A method frame F_m is a triplet consisting of a method name *m*, a program counter *i* for execution progress, both determine the program point in the invoked method and finally a register array *R* mapping register names (parameters, locals and return) to values. We adopt the same notations for registers used in Smali, as explained in the Registers subsection. Therefore, we have a set of registers for the method parameters and a set for the method local variables. Local registers content are initially undefined denoted by $\bot$. The top of the call stack represents the currently executing method's frame. Values can be either primitives or heap locations. A heap *H* map locations (we suppose an arbitrary number of unique locations) to objects *Obj* or arrays *Arr*. Objects record their class and a mapping from (class) fields to values, whereas arrays record the array type and its values. Finally, the static heap *S* is a mapping from static (class) field names to their values. Fields are annotated with their type used for initialization, to determine the default values of each primitive type (see Table 4). This annotation is omitted when it is unneeded.

The relation $\sigma \xrightarrow{m(i)} \sigma'$ models evolution of a starting configuration σ into a new σ' as the result of a computation step. $m(i)$ represents the program point, which corresponds to the instruction at a position i in a specified method m, always for the top-most method frame of the call stack in σ.

To illustrate the semantics, we present in Table 5 the semantic rules for instructions presented in Table 2.

Table 3. Semantic domains.

(Local configuration)	σ	$::= \ < C_s, H, S >$
(Call stack)	C_s	$::= \ \epsilon \mid F_m \mid C_s :: C_s$
(Method frame)	F_m	$::= \ < m, i, R >$
(Registers array)	R	$::= \ (Rg \rightarrow Val)^*$
(Registers names)	Rg	$::= \ v^* \vee p^* \vee ret$
(Heap)	H	$::= \ \epsilon \mid \ (l \rightarrow (Obj \mid Arr))^*$
(Object)	Obj	$::= \ \{|Cfn; (f_\tau \rightarrow Val)^*|\}$
(Array)	Arr	$::= \ ArrayType\,[^*\ Val$
(Static Heap)	S	$::= \ \epsilon \mid (Cfn.f_\tau \rightarrow Val)^*$
(Values)	Val	$::= \ \tau \mid l \mid \bot$
(Local register)	v	$::= \ $**string**
(Parameter register)	p	$::= \ $**string**
(Return register)	ret	$::= \ $**string**
(location)	l	$::= \ $**heap locations** $\mid$ **null**

Table 4. Default values of primitive types.

int	0
long	0|
short	0
char	$'\backslash u0000'$
byte	(byte) 0
float	0.0f
double	0.0d
object	null
boolean(int)	false (0)

Table 5. Single-threaded semantics.

R_{goto}	$m(i) = \textbf{goto } i'$ $\overline{<<m,i,R>::C_s,H,S> \xrightarrow{m(i)} <<m,i',R>::C_s,H,S>}$
$R_{mv\text{-}reg}$	$m(i) = \textbf{move } v\ Src$ $[\![Src]\!]=\mathrm{Val}$ $\overline{<<m,i,R>::C_s,H,S> \xrightarrow{m(i)} <<m,i+1,R[v\mapsto Val]>::C_s,H,S>}$
$R_{mv\text{-}sttf}$	$m(i) = \textbf{move } Cfn.f\ v$ $R(v) = Val\ \ [\![Cfn.f]\!]=S(Cfn.f)$ $\overline{<<m,i,R>::C_s,H,S> \xrightarrow{m(i)} <<m,i+1,R>::C_s,H,S[Cfn.f\mapsto Val]>}$
$R_{mv\text{-}instf}$	$m(i) = \textbf{move } v_{ref}.f'\ v$ $R(v) = Val\ \ R(v_{ref}) = l\ \ H(l) = o$ $\overline{<<m,i,R>::C_s,H,S> \xrightarrow{m(i)} <<m,i+1,R>::C_s,H[l\mapsto o[f'\mapsto Val]],S>}$
$R_{mv\text{-}cst}$	$m(i) = \textbf{move } v\ Cst$ $[\![Cst]\!]= \mathrm{Cst}$ $\overline{<<m,i,R>::C_s,H,S> \xrightarrow{m(i)} <<m,i+1,R[v\mapsto cst]>::C_s,H,S>}$
$R_{new\text{-}ins}$	$m(i) =\textbf{new-instance } v\ Cfn$ $o' = \{|Cfn; (f_\tau \mapsto 0_\tau)^*|\}\ \ l' \notin dom(H)$ $\overline{<<m,i,R>::C_s,H,S> \xrightarrow{m(i)} <<m,i+1,R[v\mapsto l']>::C_s,H[l'\mapsto o'],S>}$
$R_{b\text{-}op}$	$m(i) = \textbf{binop}_\oplus\ v\ v_1\ v_2$ $(R(v_1) \oplus R(v_2)) = Val$ $\overline{<<m,i,R>::C_s,H,S> \xrightarrow{m(i)} <<m,i+1,R[v\mapsto Val]>::C_s,H,S>}$
$R_{u\text{-}op}$	$m(i) = \textbf{unop}_\odot\ v\ v_1$ $\odot(R(v_1)) = Val$ $\overline{<<m,i,R>::C_s,H,S> \xrightarrow{m(i)} <<m,i+1,R[v\mapsto Val]>::C_s,H,S>}$
$R_{if\text{-}true}$	$m(i) =\textbf{if}_\otimes\ v_1\ v_2\ i'$ $[\![R(v_1) \otimes R(v_2)]\!] = \textbf{true}$ $\overline{<<m,i,R>::C_s,H,S> \xrightarrow{m(i)} <<m,i',R>::C_s,H,S>}$
$R_{if\text{-}false}$	$m(i) =\textbf{if}_\otimes\ v_1\ v_2\ i'$ $[\![R(v_1) \otimes R(v_2)]\!] = \textbf{false}$ $\overline{<<m,i,R>::C_s,H,S> \xrightarrow{m(i)} <<m,i+1,R>::C_s,H,S>}$
$R_{inv\text{-}st}$	$m(i) =\textbf{invoke-static } Cfn\ m'(\tau_1,...,\tau_n)ret\tau\ v_1,...,v_n$ $lookup(m'(\tau_1,..,\tau_n)ret\tau,Cfn) = m'(\tau_1,...,\tau_n)ret\tau\ loc$ $R' = \{(v_j)^{j<loc} \mapsto \bot, p_1 \mapsto R(v_0),...,p_n \mapsto R(v_n)\}$ $\overline{<<m,i,R>::C_s,H,S> \xrightarrow{m(i)} <<m',0,R'>::<m,i+1,R>::C_s,H,S>}$
$R_{inv\text{-}inst}$	$m(i) =\textbf{invoke-instance } v_{ref}\ m'(\tau_1,...,\tau_n)ret\tau\ v_1,...,v_n$ $R(v_{ref}) = l\ \ H(l) = \{|Cfn; (f_\tau \mapsto Val)^*|\}$ $lookup(m'(\tau_1,...,\tau_n)ret\tau,Cfn) = m'(\tau_1,...,\tau_n)ret\tau\ loc$ $R' = \{(v_j)^{j<loc} \mapsto \bot, p_0 \mapsto l, p_1 \mapsto R(v_0),...,p_n \mapsto R(v_n)\}$ $\overline{<<m,i,R>::C_s,H,S> \xrightarrow{m(i)} <<m',0,R'>::<m,i+1,R>::C_s,H,S>}$
$R_{ret\text{-}rtv}$	$m(i) = \textbf{return } v$ $\overline{<<m,i,R>::<m',i',R'>::C_s,H,S> \xrightarrow{m(i)} <<m',i',R'[ret\mapsto R(v)]>::C_s,H,S>}$
$R_{ret\text{-}v}$	$m(i) =\textbf{return-void}$ $\overline{<<m,i,R>::<m',i',R'>::C_s,H,S> \xrightarrow{m(i)} <<m',i',R'[ret\mapsto R(ret)]>::C_s,H,S>}$

These rules are as follows. The rule R_{goto} updates the program counter to the specified one unconditionally. Rules related to a *move* instruction from source to destination use an evaluation function $[\![-]\!]$ that evaluates a destination or a source under the current configuration σ, except for registers. In this case, for the sake of being simple, we use directly $R(v)$ always from the top-most method frame of the call stack in σ since $[\![v]\!]$ is equivalent to $R(v)$. Constants are evaluated to themselves whereas static and instance fields are evaluated based on static S and dynamic H heaps, respectively, obviously under the current configuration σ. The rule $R_{mv\text{-}reg}$ evaluates the source sub-expression and then updates the destination register content in the register array. Rules $R_{mv\text{-}insf}$ and $R_{mv\text{-}Sf}$ update instance and static field, respectively, by the content of the source register. Rule $R_{mv\text{-}cst}$ is quite straightforward. That is, after evaluating the source to constant, it updates the destination register content by the constant value.

Rule $R_{new\text{-}ins}$ creates a new object in the heap by reserving a memory with a new fresh location l, loading the class that is instantiated from and initialing its static fields, each by its default value according to Table 4. Once created, it returns the newly allocated object by pushing its heap location in a destination register v.

Rules $R_{b\text{-}op}$ and $R_{u\text{-}op}$ compute a binary or unary expression, respectively, and store the results in the destination register. Rules $R_{if\text{-}true}$ and $R_{if\text{-}false}$ models conditional jump. If the guard is evaluated to true, it branches to the targeted program counter ($R_{if\text{-}true}$), otherwise the program counter is advanced to the next instruction ($R_{if\text{-}false}$). In rules $R_{inv\text{-}st}$ and $R_{inv\text{-}dy}$, a lookup function is called to look up for the appropriate method. In the dynamic case, the method class is retrieved from the heap through object location l which is passed to the register v_{ref}. In both rules, a new method frame structure is pushed on the top of the call stack. It includes the method name, a count program set to 0 and a register array R' set as explained in the subsection Method invocation. Notice that here we increment the program counter of the caller by one to restart from the correct instruction once the callee returns.

A lookup method searches for a method matching the given method signature $(m(\tau_1, ...\tau_n) \xrightarrow{loc} \tau)$ in the given class full name and upwards to its super-class chain. Once located, it returns the method signature with the number of its local registers. We assume that the identified class and method exist in the package and class ancestry, respectively, with an array of local registers. Moreover, we admit that all verification checks are performed by the DVM. For instance it is verified that the method can be legally accessed by the class. Thus, the invoke instructions $R_{inv\text{-}st}$ and $R_{inv\text{-}dy}$ are safe to execute.

$$lookup(MtdSign, Cfn) = \begin{cases} m(\tau_1, ...\tau_n) ret\tau \ loc & if \ m \in Cfn \\ lookup(MtdSign, Cfn.Sc) & else \end{cases}$$

Rules $R_{ret\text{-}nv}$ and $R_{ret\text{-}v}$ pop the top frame from the call stack and pass on the return value from the callee back to the caller through its return register ret. Notice that, in the case of a void method, the return value must be moved to ret by the callee before the *return-void* instruction.

5. Operational Semantics for a Multi-Threaded Program

Results shown in [17] have highlighted multi-threading as a widely used feature in Android applications with 90.18% including a reference to Java/lang/Thread and 88% using monitors. An important rate that motivates us to take this feature into account in our formalization in order to develop a complete semantic.

5.1. Syntax

Here, we consider multi-threaded programs. Multi-threading semantics include single-threaded semantics for each running thread separately. Threads in the same DVM interact and synchronize using shared objects and monitors associated with these objects. In order to give a full account of Java concurrency, we consider instructions related to this aspect. We define macro-instructions that cover methods of the Java Thread API [40] which are *start* for thread spawning and *join* for joining a referenced thread. We also define macro-instructions that cover several methods of the Java Object API [41] related to thread signaling such as *notify*, *notifyAll* and to synchronization such as *wait*. We also give the semantics of Dalvik instructions related to threads synchronization and monitors with the instructions *monitor-enter* and *monitor-exit*. All instructions syntax are illustrated in Table 6.

Table 6. Smali$^+$: concurrent instructions.

Inst	::=	**start** v_{ref}	(start the thread in v_{ref})
	|	**monitor-enter** v_{ref}	(acquire the monitor for object in v)
	|	**monitor-exit** v_{ref}	(release the monitor for object in v)
	|	**join** v_{ref}	(join the thread in v_{ref})
	|	**wait** v_{ref}	(release object's monitor in v_{ref} and suspend current thread)
	|	**notify** v_{ref}	(notify one thread from those waiting on object's monitor in v_{ref})
	|	**notifyAll** v_{ref}	(notify all threads waiting on object's monitor in v_{ref})

5.2. Semantics

An overall configuration $\Sigma =< C_s, S_{rbl}, H, S >$ models the full state of an Android application in its low-level implementation. It presents a multi-threading program configuration including as first attribute a running thread's call stack C_s, a set of runnable threads S_{rbl}, a heap H and a static heap S.

- Each thread in the program has a call stack C_s for methods being invoked, their arguments and local variables, with the same syntax used in Table 3.
- S_{rbl} is a set of pending threads. Each thread is presented by its call stack for method invoked information, plus a special register p_0 holding the thread reference. Threads in this set are in a "runnable" state (i.e., waiting to be selected by the scheduler).
- H and S are dynamic and static heaps which are shared between all threads in the program and have the same semantics domain used for the single-threaded program in Table 3.

A new semantic domain for multi-threaded program is provided in Table 7. Some changes are applied to the object definition. It includes a new fields *acq* which indicates if the object's monitor is acquired by another thread. If this is the case, *acq* will contain this thread's reference, otherwise it will contain an undefined value $\perp$ since an object cannot be reserved by more than one thread at once, at a given time. S_{blck} is a set of blocked threads waiting for the object's monitor to be released. S_{wait} is a set of threads pending notification (threads that executed the *wait* instruction). The initial state of a new instance object, in a multi-threading context, will be initialized as seen in the single-threaded environment (with default values). New attributes are initialized as follows:

- $acq \mapsto \perp$ initialized to an undefined value, which means that initially the object is in a free state and could be acquired by a given thread.
- $S_{blck} \mapsto \varnothing$, an empty set of blocked threads, which means that initially there is no thread waiting for the monitor to be released.
- $S_{wait} \mapsto \varnothing$, an empty set of waiting to be notified threads.

A class Cl is a Thread class if and only if it is an instance of a Thread class ($\perp$=Thread), which means that its super class Sc is either the Thread top class path ($Cfn = $ LJava/lang/Thread) or another class that it is extended from this class. Each thread object has a Boolean finished field indicating whether the thread has completed its execution or not, a mapping from a group of threads to a set of threads call stacks, it contains a set of threads waiting to join this thread and an attribute called *state* indicating the current state of the thread. Each thread has a run method. Thread attributes are initialized as follows:

- *finished* $\mapsto$ *false*.

- $S_{join} \mapsto \varnothing$, an empty set of join threads, which means that initially there is no thread waiting to join the current thread.
- $state = \perp$.

Table 7. Semantic domains for a multi-threaded program.

(Global configuration)	Σ	$::= \; < C_s, S_{rbl}, H, S >$	
(Set of runnable threads)	S_{rbl}	$::= \; \varnothing \mid C_s \mid \{S_{rbl}, S_{rbl}\}$	
(Object)	Obj	$::= \; \{\mid Cfn; (f_\tau \to Val)^*; \; acq \mapsto Val; \; S_{blck} \mapsto S_b; \; S_{wait} \mapsto S_w \mid\}$	
(A thread Object)	th	$::= \; \{\mid Cfn; (f_\tau \to Val)^*; \; finished \mapsto booelan; \; S_{join} \mapsto S_j \mid\}$	
(Set of blocked threads)	S_b	$::= \; \varnothing \mid C_s \mid \{S_b, S_b\}$	
(Set of waiting threads)	S_w	$::= \; \varnothing \mid C_s \mid \{S_w, S_w\}$	
(Set of join threads)	S_j	$::= \; \varnothing \mid C_s \mid \{S_j, S_j\}$	
(Acquiring field)	acq	$:: \; f$	(field name)
(finished field)	$finished$	$:: \; f$	(field name)
(Groups names)	$S_{wait}, S_{blck}, S_{join}$	$::= \mathbf{String}$	

Table 8 provides the semantics of spawning and scheduling threads. Rule R_{start} starts a new thread, which reference is stored in the register v_{ref}. It internally calls the referenced thread's $run()$ method that will be executed in this thread separately, once selected. Therefore, a $lookup()$ procedure for its run method is performed and a separate call stack for a new thread is created with one frame comprising all information about the thread's $run()$ method returned by the $lookup()$ function. This thread moves to a "runnable" state in S_{rbl}. When it gets a chance to execute, its target run() method will be executed. The actual execution of the launched thread will be managed with the rule R_{select}. Notice that, as expressed by the rule R_{start}, the reference of the launched thread is always stored in the register p_0 and we assume that it will remain there for all semantics rules and for all method's frames in the thread's call stack.

Table 8. Multi-threaded semantics: scheduling.

$$R_{start} \quad \frac{\begin{array}{c} m(i) = \mathbf{start}\ v_{ref} \\ R(v_{ref}) = l \quad H(l) = \{\mid Cfn; (f_\tau \to Val)^*; finished \mapsto false; \; S_{join} \mapsto S_j \mid\} \\ lookup(run()V, Cfn) = run()V\ loc \\ R' = \{(v_j)^{j < loc} \mapsto \perp, \mathbf{p_0} \mapsto \mathbf{1}\} \quad F_m = <\mathbf{@run}, 0, R' > \end{array}}{<<<m,i,R>::C_s,S_{rbl},H,S> \xrightarrow{m(i)} <<m,i+1,R>::C_s,S_{rbl}\cup\{F_m\},H,S>}$$

$$R_{select} \quad \frac{\begin{array}{c} selectFrom(S_{rbl}) = [F_m :: C_s, t_s]\ F_m = < m, i, R > \\ R(p_o) = l \quad H(l) = \{\mid Cfn; (f_\tau \to Val)^*; finished \mapsto false; state \mapsto\ -\ S_{join} \mapsto S_j \mid\} \\ th' = \{\mid Cfn; (f_\tau \to Val)^*; finished \mapsto false; state \mapsto Running(t_s)\ S_{join} \mapsto S_j \mid\} \end{array}}{<\epsilon,S_{rbl},H,S> \xrightarrow{\tau} <F_m::C_s,S_{rbl}\setminus\{F_m::C_s\},H[l \mapsto th'],S>}$$

$$R_{stop} \quad \frac{\begin{array}{c} R(p_o) = l \quad H(l) = \{\mid Cfn; (f_\tau \to Val)^*; finished \mapsto false; state \mapsto Running(t_s)\ S_{join} \mapsto S_j \mid\} \\ clock() > t_s \end{array}}{<<m,i,R>::C_s,S_{rbl},H\,S>_\mathbf{M} \xrightarrow{\tau} <\epsilon,S_{rbl}\cup\{C_s\},H,S>_\mathbf{M}}$$

Rules R_{select} and R_{stop} manage threads scheduling. Rule R_{select} selects from S_{rbl} one thread to be executed for a time slice t_s. The selected thread's state will be updated to a "Running(t_s)" state. The thread's call stack will be removed from the runnable set and placed at the first position of configuration Σ to start execution. The $select(S_{rbl})$ function will be based on a CFS scheduler's algorithm for scheduling threads in S_{rbl}. It takes into account the thread's nice values and returns the selected thread's local state presented in its current call stack as well as the time slice allocated to it for execution.

Rule R_{stop} stops, in a monitoring mode (i.e., a mode that monitors the execution time given to each thread), a thread whose allocated time slice to execute a task has expired. We model the timing aspect in our formalism by the function *clock*() which represents the scheduler timer to control running threads.

Synchronization in *Dalvik* is modeled by the use of monitors with instructions *monitor-enter* and *monitor-exit*. That actually corresponds to the *synchronized* keyword in Java. A monitor is attached to an object and could be acquired and released by threads.

The semantics of these two instructions must fulfill two conditions. The first is related to the mutual exclusive access to shared objects in the heap by different threads. The second relates to the cooperation between these threads. Cooperation is modeled by a set of threads waiting for notification when the object is released by another thread. The sole thread running and owning the monitor is in a critical section. Table 9 presents rules related to synchronization. *Monitor-enter* semantics represent a thread trying to access the critical section by acquiring monitor for the object, whose reference is stored in a register v_{ref}. It first checks if the object is acquired by any other thread. If this is the case, the current thread will be blocked (mutual exclusive access condition) and added to the object blocking set S_{blck} to join other threads (if any) with the same situation (cooperation condition). This case is modeled by the rule R_{block}). Otherwise, the current thread can take ownership of the monitor. The *acq* attribute is then updated with this thread's reference. This thread could resume its execution in the critical section. This case is modeled with the rule $R_{acq-mnt}$.

Table 9. Multi-threaded semantics: synchronization.

$$R_{Acq-mntr} \quad \frac{m(i) = \textbf{monitor-enter } v_{ref} \qquad R(v_{ref}) = 1 \quad H(l) = \{|Cfn; (f_\tau \to Val)^*; acq \mapsto \bot; S_{blck} \mapsto S_b; S_{wait} \mapsto S_w|\} \qquad o' = \{|Cfn; (f_\tau \to Val)^*; acq \mapsto R(p_0); S_{blck} \mapsto S_b; S_{wait} \mapsto S_w|\}}{<<m,i,R>::C_s,S_{rbl},H,S> \xrightarrow{m(i)} <<m,i+1,R>::C_s,S_{rbl},H[l\mapsto o'],S>}$$

$$R_{block} \quad \frac{m(i) = \textbf{monitor-enter } v_{ref} \qquad R(v_{ref}) = 1 \quad H(l) = \{|Cfn; (f_\tau \to Val)^*; acq \mapsto l'; S_{blck} \mapsto S_b; S_{wait} \mapsto S_w|\} \qquad o' = \{|Cfn; (f_\tau \to Val)^*; acq \mapsto l'; S'_{blck} \mapsto S_b \cup \{< m,i,R >:: C_s\}; S_{wait} \mapsto S_w|\}}{<<m,i,R>::C_s,S_{rbl},H,S> \xrightarrow{m(i)} <<m,i,R>::C_s,S_{rbl},H[l\mapsto o'],S>}$$

$$R_{Rls-mntr} \quad \frac{m(i) = \textbf{monitor-exit } v_{ref} \qquad R(p_0) = l' \quad R(v_{ref}) = 1 \quad H(l) = \{|Cfn; (f_\tau \to Val)^*; acq \mapsto l'; S_{blck} \mapsto S_b; S_{wait} \mapsto S_w|\} \qquad o' = \{|Cfn; (f_\tau \to Val)^*; acq \mapsto \bot; S'_{blck} \mapsto \varnothing; S_{wait} \mapsto S_w|\}}{<<m,i,R>::C_s,S_{rbl},H,S> \xrightarrow{m(i)} <<m,i+1,R>::C_s,S_{rbl}\cup S_b,H[l\mapsto o'],S>}$$

$$R_{wait} \quad \frac{m(i) = \textbf{wait } v_{ref} \qquad R(p_0) = l' \; R(v_{ref}) = 1 \quad H(l) == \{|Cfn; (f_\tau \to Val)^*; acq \mapsto l'; S_{blck} \mapsto S_b; S_{wait} \mapsto S_w|\} \qquad o' = \{|Cfn; (f_\tau \to Val)^*; acq \mapsto \bot; S_{blck} \mapsto \varnothing; S'_{wait} \mapsto S_w \cup \{< m,i,R >:: C_s\} |\}}{<m,i,R>::C_s,S_{rbl},H,S> \xrightarrow{m(i)} <<m,i,R>::C_s,S_{rbl}\cup S_b,H[l\mapsto o'],S>}$$

Monitor-exit semantics represents a thread that reaches the end of the critical section by releasing the owned monitor for another thread to take ownership, which perfectly fulfills the cooperation condition. Rule $R_{Rls-mntr}$ provides this semantics, the current thread must first own this object's monitor, once this condition is satisfied, the *acq* attribute is updated to an undefined value (object is free). Then, all waiting threads in S_{blck} are removed to the runnable set S_{rbl}. It is up to the scheduler to select which thread to execute (there is no ordering among the blocked threads).

A thread could voluntarily give up ownership of the monitor before reaching the end of the critical section by calling the *wait*() method or by executing the *wait* instruction. This thread releases ownership of this monitor and remains in a waiting state (i.e., suspended or inactive until be notified by another thread). Rule R_{wait} provides the semantics of wait instruction. The calling thread must own this object's monitor (i.e., must executing *wait* from inside a synchronized block) then relinquish it. Once the monitor associated with this object is released, the current thread is placed in the wait set for this object.

Table 10 presents rules R_{notify} and $R_{notifyAll}$ expressing the signaling mechanism. Rule R_{notify} represents the semantics for waking up a single thread that is waiting for this object's monitor in the waiting set S_{wait}. One thread among the set will be chosen randomly by the function $random()$. This thread will be moved from the waiting set to the runnable set to be selected later on by the scheduler and then processed. The rule $R_{notifyAll}$ is similar to the rule R_{notify}, with the exception that it wakes all threads in the waiting set, which ones will be moved to the runnable set S_{rbl}. Notice that, rules R_{notify} and $R_{notifyAll}$ release in addition to waiting thread(s) set S_{wait} all blocked threads in S_{blck}. The two sets have the same privileges with regards to acquiring monitor. In other words, waiting threads have no precedence over potentially blocked threads that also want to synchronize on this object.

Table 10. Multi-threaded semantics: signaling.

	$m(i) = \textbf{notify}\ v_{ref}$
R_{notify}	$\dfrac{R(p_0) = l \quad R(v_{ref}) = l' \quad H(l') = \{\lvert Cfn; (f_\tau \to Val)^*; acq \mapsto l; S_{blck} \mapsto S_b; S_{wait} \mapsto S_w \rvert\} \quad random(S_w) = C'_s \quad o' = \{\lvert Cfn; (f_\tau \to Val)^*; acq \mapsto \perp; S'_{blck} \mapsto \varnothing; S'_{wait} \mapsto S_w \setminus \{C'_s\} \rvert\}}{<<m,i,R>::C_s,S_{rbl},H,S> \xrightarrow{m(i)} <<m,i+1,R>::C_s,S_{rbl}\cup\{C'_s\}\cup S_b,H[l'\mapsto o'],S>}$
	$m(i) = \textbf{notifyAll}\ v_{ref}$
$R_{notifyAll}$	$\dfrac{R(p_0) = l \quad R(v_{ref}) = l' \quad H(l') = \{\lvert Cfn; (f_\tau \to Val)^*; acq \mapsto l; S_{blck} \mapsto S_b; S_{wait} \mapsto S_w \rvert\} \quad o' = \{\lvert Cfn; (f_\tau \to Val)^*; acq \mapsto \perp; S'_{blck} \mapsto \varnothing; S'_{wait} \mapsto \varnothing \rvert\}}{<<m,i,R>:::C_s,S_{rbl},H,S> \xrightarrow{m(i)} <<m,i+1,R>::C_s,S_{rbl}\cup S_w\cup S_b,H[l'\mapsto o'],S>}$

Table 11 presents semantics of finishing thread and joining instructions. Rules $R_{Join\text{-}exec}$ and $R_{Join\text{-}wait}$ check if the joined thread has finished its execution, if so, the current thread resumes execution ($R_{Join\text{-}exec}$). Otherwise, the rule $R_{Join\text{-}wait}$ is applied. The current running thread is removed into S_{join} for threads waiting for the same thread to complete its execution (no release by the monitor of the object is acquired by the running thread here). The rule R_{finish} ensures that when a thread completes its execution (i.e., its run() method returns) and releases all waiting threads in S_{join} by moving them to the runnable set S_{rbl}.

Table 11. Multi-threaded semantics: join.

	$m(i) = \textbf{join}\ v_{ref}$
$R_{Join\text{-}exec}$	$\dfrac{R(v_{ref}) = l \quad H(l) = \{\lvert Cfn; (f_\tau \to Val)^*; finished \mapsto true;\ S_{join} \mapsto S_j \rvert\}}{<<m,i,R>::C_s,S_{rbl},H,S> \xrightarrow{m(i)} <<m,i+1,R>::C_s,S_{rbl},H,S>}$
	$m(i) = \textbf{join}\ v_{ref}$
$R_{join\text{-}wait}$	$\dfrac{R(v_{ref}) = l \quad H(l) = \{\lvert Cfn; (f_\tau \to Val)^*; finished \mapsto false;\ S_{join} \mapsto S_j \rvert\} \quad o' = \{\lvert Cfn; (f_\tau \to Val)^*; finished \mapsto false;\ S'_{join} \mapsto S_j \cup \{< m,i,R >:: C_s\} \rvert\}}{<<m,i,R>::C_s,S_{rbl},H,S> \xrightarrow{m(i)} <<m,i,R>::C_s,S_{rbl},H[l\mapsto o'],S>}$
	$@run(i) = \textbf{return-void}$
R_{finish}	$\dfrac{R(p_0) = l \quad H(l) = \{\lvert Cfn; (f_\tau \to Val)^*; finished \mapsto false;\ S_{join} \mapsto S_j \rvert\} \quad o' = \{\lvert Cfn; (f_\tau \to Val)^*; finished \mapsto true;\ S'_{join} \mapsto \varnothing \rvert\}}{<<@run,i,R>::\epsilon,S_{rbl},H,S> \xrightarrow{@run(i)} <\epsilon,S_{rbl}\cup S_j,H[l\mapsto o'],S>}$

6. Practical Aspects

We give, hereafter, some practical aspects of Smali+ through an example. For the sake of simplicity and due to the space limitation, we only present an illustration of a single-threaded program in Smali$^+$ that includes various important instructions such as method call, return, static and instance field update, etc. As shown in Table 12, the program is sequential and consists of two classes $c1$ and $c2$ belonging to the same package called p. Figure 3 shows the initial configuration. We show in detail, through this example, how the rules are applied and how the configuration evolves in every step. Each rule is followed by the resulting configuration.

Table 12. Smali$^+$ program.

.class public $Lp/c2$ **.super** c1 {	**.class public** $Lp/c1$ **.super** $\bot$ {
.field public x: int	**.field** public a: LJava/lang/String
.field public y: char	**.field public** b: int
.method public static m_1()V **.locals** 3 {	**.field private final** c: char
...	**.method public static** m_2(int,char)char **.locals** 2 {
5 **move** v_1 30	... ...
6 **goto** 10	
... ...	
10 **invoke-static** $Lp/c1$ m_2(int,char)char v_0,v_1	
11 **move** c2.x v_0	
12 **new-instance** v_2 Lp/c1	... ...
13 **move** $v_2.b$ v_1	18 **return** v_1
... ...	}
}	

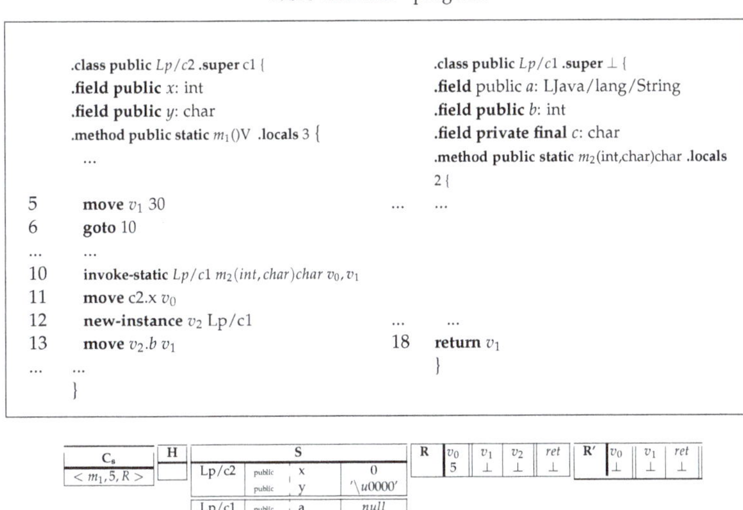

Figure 3. Initial configuration.

The first table corresponds to the call stack C_s, which is the current method frame. The second table corresponds to an empty heap H and the last two tables correspond to the register arrays for methods m_1 and m_2, respectively.

The first Smali$^+$ instruction to execute is the move instruction labeled with 5. It is a constant displacement, so the rule $R_{mv\text{-}cst}$ applies. Since constants are evaluated to themselves, the register v_1 for m_1 locals registers is updated by the constant value and the program counter is incremented.

$$R_{mv\text{-}cst} \quad \frac{m_1(5) = \textbf{move}\ v_1\ 30 \qquad [\![30]\!] = 30}{<<m_1,5,R>::C_s,H,S> \xrightarrow{m_1(5)} <<m_1,6,R[v_1 \mapsto 30]>::C_s,H,S>}$$

The next instruction corresponds to the unconditional jump *goto*. The rule R_{goto} so applies to update the program counter by the instruction labeled with 10.

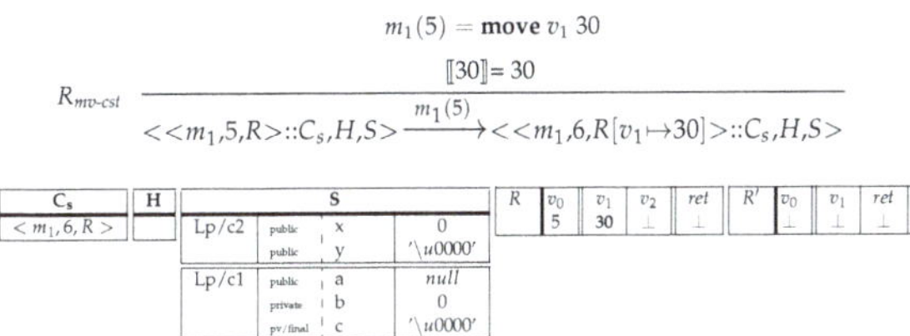

$$R_{goto} \quad \frac{m_1(6) = \textbf{goto}\ 10}{<<m_1,6,R>::C_s,H,S> \xrightarrow{m_1(6)} <<m_1,10,R>::C_s,H,S>}$$

$m_1(10)$ is an invocation of a static method. Rule $R_{inv\text{-}st}$ so applies. A new frame for the called method is pushed on top of C_s and the counter program in the caller method frame is incremented.

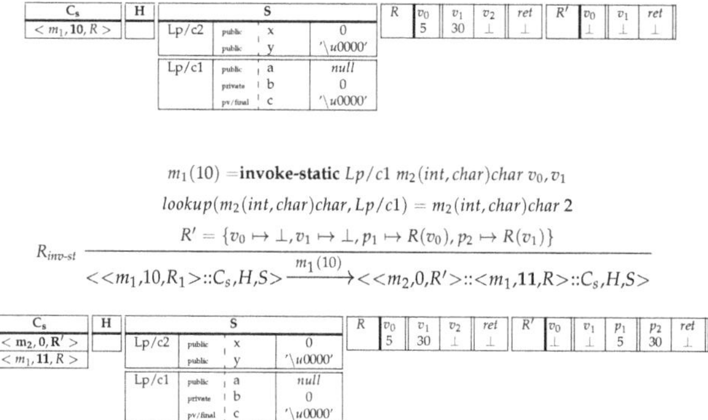

$$m_1(10) = \textbf{invoke-static } Lp/c1 \; m_2(int, char)char \; v_0, v_1$$

$$lookup(m_2(int,char)char, Lp/c1) = m_2(int,char)char \; 2$$

$$R_{inv\text{-}st} \; \frac{R' = \{v_0 \mapsto \bot, v_1 \mapsto \bot, p_1 \mapsto R(v_0), p_2 \mapsto R(v_1)\}}{<<m_1,10,R_1>::C_s,H,S> \xrightarrow{m_1(10)} <<m_2,0,R'>::<m_1,11,R>::C_s,H,S>}$$

After some execution steps, we suppose that the register v_1 in m_2 is updated by a new value "CA" and the current instruction to execute is labeled with 18 in m_2.

The instruction $m_2(18)$ is a return from a non-void method m_2, so the rule $R_{ret\text{-}nv}$ applies. The top frame of C_s is popped and the return value is passed from the callee back to the caller through its return register *ret*.

$$R_{ret\text{-}nv} \; \frac{m_2(18) = \textbf{return } v_1}{<<m_2,18,R'>::<m_1,11,R>::C_s,H,S> \xrightarrow{m_2(18)} <<m_1,11,R[ret \mapsto R'(v_1)]>::C_s,H,S>}$$

The instruction $m_1(11)$ is a static field update. So the rule $R_{mv\text{-}stf}$ so applies to update the indicated field in the static heap S by the register v_0 content.

$$m_1(11) = \textbf{move } Lp/c2.x \; v_0$$

$$R_{mv\text{-}stf} \; \frac{R(v_0) = 5 \quad [\![Lp/c2.x]\!] = S(Lp/c2.x)}{<<m_1,11,R>::C_s,H,S> \xrightarrow{m_1(11)} <<m_1,12,R>::C_s,H,S[Lp/c2.x \mapsto 5]>}$$

The instruction $m_1(12)$ corresponds to an object creation. The rule $R_{new\text{-}instance}$ so applies to create a new instance from the class $c1$ in the heap H and all fields are initialized according to their types.

$$m_1(12) = \textbf{new-instance } v_2 \; Lp/c1$$

$$R_{new\text{-}ins} \; \frac{o' = \{|Lp/c1; (a \mapsto null, b \mapsto \backslash u0000', c \mapsto \backslash u0000')\} \quad l' \notin dom(H)}{<<m_1,12,R>::C_s,H,S> \xrightarrow{m_1(12)} <<m_1,13,R[v_2 \mapsto l']>::C_s,H[l' \mapsto o'],S>}$$

The instruction $m_1(13)$ is an instance field update. So the rule $R_{mv\text{-}instf}$ applies. The register v_2 holds the instance location o' in H. The instance field in o' is updated with the source register v_1 content.

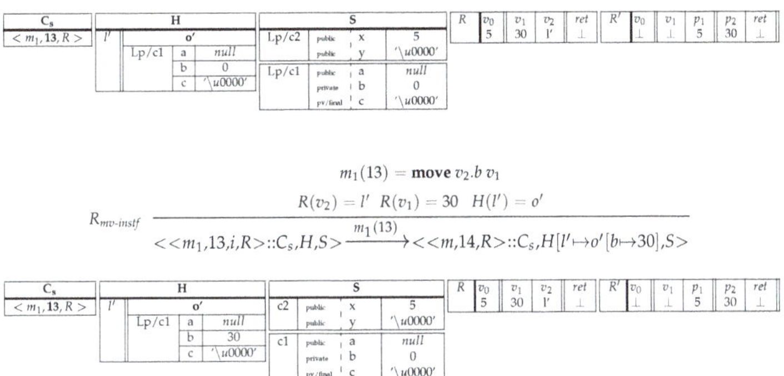

7. Discussion

So far, we have proposed a formal language for Android programs called $Smali^+$. Presented in a BNF notation, $Smali^+$ is a simple language that remains faithful to the original Smali notations and the .Smali file structure. It contains 12 generalized instructions from 218 Dalvik instructions [39] and some macros instructions modeling concurrency aspect. These 12 instructions were selected carefully to highlight Dalvik's characteristics, such as register-based architecture, assembly-like code for Smali, methods invocations, monitors, etc. Macro instructions were used for the sake of simplification as well as to model multi-threading in Android. All the important API methods that affect a thread life-cycle were considered in $Smali^+$ semantics.

Another important feature that lacks so far in Android application semantics is thread scheduling. This important aspect, in general, consists in picking a thread for execution and allocating an execution time to it, depending on its priority, before selecting a new thread to execute and switching the context. Android applications including their threads adhere to the Linux execution environment. So, threads are scheduled using the standard scheduler of the Linux kernel, known as a *completely fair scheduler* (CFS). On Linux, the thread priority is called a "nice value". A low nice value corresponds to a high priority and vice versa. In Android, a Linux thread has niceness values in the range of -20 (most prioritized) to 19 (least prioritized), with a default niceness of 0 [42]. We exhibited in this work two rules related to scheduling feature in Android, R_{select} and R_{stop}. In the first rule, we presented a function $select()$ that plays the same role as the CFS, meaning it selects from runnable threads the most prioritized thread based on nice values comparison and allocates to it an amount of time for execution. The second rule stops a thread when the allocated time expires, prior to picking a new one through R_{select}. We mean by "monitoring mode" mentioned in threads scheduling, a monitor that is based on the CFS algorithm that monitors each thread for each task executed, and we suppose that each rule in the concurrent context is executing under a monitoring mode. This mode was presented just for R_{stop} and omitted in other rules for simplification reasons.

The operational semantics are mainly created to secure Android applications. In fact, we intend to use these semantics in an upcoming work to check a number of security proprieties to protect users from rogue applications. Our ultimate goal is to formally reinforce security policies on Android applications. That is to say, starting from a $Smali^+$ program and a formal specification of a security policy, we automatically generate a new equivalent secure version of the original program that respects the security policy. Formally, the approach takes, as input, a $Smali^+$ program P and a formal specification of a security policy ϕ and generates, as output, a new version P' that respects ϕ. The new version of the program preserves all the behavior of the original version, except in cases where the security policy is on the verge of being violated. This is equivalent to saying that the traces of P' are the intersection of traces accepted by ϕ and traces of P. It is formally modeled by (1).

$$P' = P \cap \phi \tag{1}$$

Security policies will be enforced through a program-rewriting approach that combines static and dynamic approaches. It rewrites the program statically, according to a given security property, then generates a new executable version that satisfies this property. Security modifications or tests are added at well-calculated points in the program to force the latter to conform to the security property during execution. In other words, the untrusted code will be transformed into a self-monitoring code that will be exploded at specific points in the program. The rewritten version should be equivalent but more restrictive than the original so that it will be able to avoid potentially dangerous operations before they occur.

Reinforced security properties will obviously be specific to malware and attacks threatening Android applications, such as sensitive information leakage, which could be SMS contents, call logs, contact information or geographical location or Android financial malware, which exploit the premium services to incur financial loss to the user for the benefit of the attacker, for example, by calling or texting to premium-rate numbers without the user's consent and privilege escalation attacks [43]. Therefore, all mediums that could be exploited for this kind of malware, such as Internet access, system services access including SMS, contact, telephony, Bluetooth, Global Positioning System (GPS) as well as APIs resulted from inter-application communication, will be checked through security policies. Such APIs will be easily located in Smali$^+$, since it provides for each invocation the class fully qualified name.

8. Conclusions

In this paper, we have proposed a formal operational semantics for Smali, an assembly-like code generated form reverse engineering Android applications. We called the new formal language Smali$^+$. Smali$^+$ covers the semantics of a large subset of the main Dalvik instructions as well as many important aspects related to multi-threading programming which are rarely considered in the state-of-the-art works of Android applications. This formal model is meant to be an environment to run formal verification of applications. Broader work consisting in techniques to reinforce the security of Android applications using this formalism is currently underway. We are deeply convinced that this will be of great help in analyzing the security of Android applications and verifying their hidden functions affecting users' privacy as well as protecting users from malicious actions.

Author Contributions: Conceptualization, M.Z., J.F. and M.M.; methodology, M.Z., M.M. and J.F; validation, J.F., M.M. and E.P.; formal analysis, M.Z., M.M. and J.F.; investigation, M.Z., J.F., M.M. and E.P.; resources, M.Z., J.F., M.M. and E.P.; writing–original draft preparation, M.Z. and J.F.; writing–review and editing, M.Z. and J.F.; supervision, M.M., J.F. and E.P.; project administration, M.M.; funding acquisition, M.M. All authors have read and agreed to the published version of the manuscript.

Funding: This research was funded by the Natural Sciences and Engineering Research Council of Canada (NSERC).

Conflicts of Interest: The authors declare no conflict of interest.

References

1. IDC Corporation. Smartphone Market Share. Available online: https://www.idc.com/promo/smartphone-market-share/os (accessed on 19 February 2020).
2. Zhou, Y.; Jiang, X. Dissecting Android Malware: Characterization and Evolution. In Proceedings of the 2012 IEEE Symposium on Security and Privacy, San Francisco, CA, USA, 20–23 May 2012. [CrossRef]
3. Sergiu Gatlan. Anubis Android Trojan Spotted with Almost Functional Ransomware Module. Available online: https://www.bleepingcomputer.com/news/security/anubis-android-trojan-spotted-with-almost-functional-ransomware-module/ (accessed on 20 February 2020).
4. Barrett, L. SMS-Sending Trojan Targets Android Smartphones. Available online: https://www.esecurityplanet.com/trends/article.php/3898041/SMSSending-Trojan-Targets-Android-Smartphones.htm/ (accessed on 2 January 2020).

5. Collier, N. New Android Trojan Malware Discovered in Google Play. Available online: https://blog.malwarebytes.com/cybercrime/2017/11/new-trojan-malware-discovered-google-play// (accessed on 2 January 2020).

6. F-Secure. Trojan:Android/GGTracker Available online: https://www.f-secure.com/v-descs/trojan_android_ggtracker.shtml (accessed on 2 January 2020).

7. Arzt, S.; Rasthofer, S.; Fritz, C.; Bodden, E.; Bartel, A.; Klein, J.; Le Traon, Y.; Octeau, D.; McDaniel, P. FlowDroid: Precise Context, Flow, Field, Object-sensitive and Lifecycle-aware Taint Analysis for Android Apps. *SIGPLAN Not.* **2014**, *49*, 259–269. [CrossRef]

8. Felt, A.P.; Chin, E.; Hanna, S.; Song, D.; Wagner, D. Android Permissions Demystified. In Proceedings of the 18th ACM Conference on Computer and Communications Security, New York, NY, USA, October 2011; doi:10.1145/2046707.2046779. [CrossRef]

9. Davis, B.; Sanders, B.; Khodaverdian, A.; Chen, H. I-arm-droid: A rewriting framework for in-app reference monitors for android applications. In Proceedings of the Mobile Security Technologies 2012, San Francisco, CA, USA, May 2012. Available online: http://citeseerx.ist.psu.edu/viewdoc/download?doi=10.1.1.298.7191&rep=rep1&type=pdf (accessed on 2 January 2020).

10. Xu, R.; Saïdi, H.; Anderson, R.J. Aurasium: Practical Policy Enforcement for Android Applications. In Proceedings of the 21th USENIX Security Symposium, Bellevue, WA, USA, 8–10 August 2012; pp. 539–552.

11. Jeon, J.; Micinski, K.K. *SymDroid : Symbolic Execution for Dalvik*; CS-TR-5022; University of Maryland: College Park, MD, USA, July 2012. Available online: http://www.cs.tufts.edu/~jfoster/papers/symdroid.pdf (accessed on 4 January 2020).

12. Apktool. A Tool for Reverse Engineering Android Apk Files. Available online: https://ibotpeaches.github.io/Apktool/ (accessed on 19 February 2019).

13. Na, G.; Lim, J.; Kim, K.; Yi, J.H. Comparative Analysis of Mobile App Reverse Engineering Methods on Dalvik and ART. *J. Internet Serv. Inf. Secur.* **2016**, *6*, 27–39.

14. El-Zawawy, M.A. An Operational Semantics for Android Applications. In Proceedings of the Computational Science and Its Applications - ICCSA 2016 - 16th International Conference, Beijing, China, 4–7 July 2016; pp. 100–114.

15. Payet, E.; Spoto, F. An Operational Semantics for Android Activities. Available online: https://doi.org/10.1145/2543728.2543738 (accessed on 5 December 2019).

16. Wognsen, E.; Karlsen, S. Static Analysis of Dalvik Bytecode and Reflection in Android. Master's Thesis, Department of Computer Science, Aalborg University, Aalborg, Denmark, 6 June 2012. Available online: https://projekter.aau.dk/projekter/files/63640573/rapport.pdf (accessed on 10 December 2019).

17. Wognsen, E.; Søndberg Karlsen, H.; Chr. Olesen, M.; Hansen, R. Formalisation and analysis of Dalvik bytecode. *Sci. Comput. Program.* **2014**, *92*, 25–55. [CrossRef]

18. Cousot, P.; Cousot, R. Abstract Interpretation: A Unified Lattice Model for Static Analysis of Programs by Construction or Approximation of Fixpoints. In Proceedings of the 4th ACM SIGACT-SIGPLAN Symposium on Principles of Programming Languages, New York, NY, USA, January 1977; pp. 238–252. [CrossRef]

19. Payet, E.; Spoto, F. Static Analysis of Android Programs. *Inf. Softw. Technol.* **2012**, *54*, 1192–1201. [CrossRef]

20. Gunadi, H. Formal Certification of Non-interferent Android Bytecode (DEX Bytecode). In proceedings of the 2015 20th International Conference on Engineering of Complex Computer Systems ICECCS, Gold Coast, Australia, 9–12 December 2015; pp. 202–205.

21. Gunadi, H.; Tiu, A.; Gore, R. Formal Certification of Android Bytecode. *arXiv* **2015**, arXiv:1504.01842v5. Available online: https://arxiv.org/abs/1504.01842 (accessed on 19 February 2020).

22. Barthe, G.; Pichardie, D.; Rezk, T. A certified lightweight non-interference Java bytecode verifier. *Math. Struct. Comput. Sci.* **2013**, *23*, 1032–1081. [CrossRef]

23. Maiya, P.; Kanade, A.; Majumdar, R. Race detection for Android applications. In Proceedings of the ACM SIGPLAN Conference on Programming Language Design and Implementation, PLDI '14, Edinburgh, UK, 9–11 June 2014; pp. 316–325.

24. Kanade, A. Chapter Seven - Event-Based Concurrency: Applications, Abstractions, and Analyses. *Adv. Comput.* **2019**, *112*, 379–412.

25. Bouajjani, A.; Emmi, M.; Enea, C.; Ozkan, B.K.; Tasiran, S. Verifying Robustness of Event-Driven Asynchronous Programs Against Concurrency. In Proceedings of the Programming Languages and Systems 26th European Symposium on Programming, ESOP 2017, Held as Part of the European Joint Conferences on Theory and Practice of Software, Uppsala, Sweden, 22–29 April 2017; pp. 170–200.

26. Calzavara, S.; Grishchenko, I.; Koutsos, A.; Maffei, M. A Sound Flow-Sensitive Heap Abstraction for the Static Analysis of Android Applications. *arXiv* **2017**, arXiv:1705.10482v2. Avaiable online: https://arxiv.org/pdf/1705.10482.pdf (accessed on 15 December 2019).

27. Calzavara, S.; Grishchenko, I.; Maffei, M. HornDroid: Practical and Sound Static Analysis of Android Applications by SMT Solving. In Proceedings of the 2016 IEEE European Symposium on Security and Privacy (EuroSP), aarbrucken, Germany, 21–24 March 2016. [CrossRef]

28. Chaudhuri, A. Language-based security on Android. In Proceedings of the 2009 Workshop on Programming Languages and Analysis for Security, Dublin, Ireland, 15–21 June 2009. [CrossRef]

29. Chen, T.; He, J.; Song, F.; Wang, G.; Wu, Z.; Yan, J. Android Stack Machine. Computer Aided Verification. In Proceedings of the 30th International Conference, CAV 2018, Held as Part of the Federated Logic Conference, Oxford, UK, 14–17 July 2018; pp. 487–504. [CrossRef]

30. He, J.; Chen, T.; Wang, P.; Wu, Z.; Yan, J. Android Multitasking Mechanism: Formal Semantics and Static Analysis of Apps. In Proceedings of the Programming Languages and Systems - 17th Asian Symposium, Nusa Dua, Bali, Indonesia, 1–4 December 2019; pp. 291–312. [CrossRef]

31. Bagheri, H.; Kang, E.; Malek, S.; Jackson, D. Detection of Design Flaws in the Android Permission Protocol Through Bounded Verification. In Proceedings of the FM 2015: Formal Methods - 20th International Symposium, Oslo, Norway, 24–26 June 2015; pp. 73–89. [CrossRef]

32. Ren, L.; Chang, R.; Yin, Q.; Man, Y. A Formal Android Permission Model Based on the B Method. In Proceedings of the Security, Privacy, and Anonymity in Computation, Communication, and Storage 10th International Conference, Guangzhou, China, 12–15 December 2017; pp. 381–394. [CrossRef]

33. Khan, W.; Kamran, M.; Ahmad, A.; Khan, F.A.; Derhab, A. Formal Analysis of Language-Based Android Security Using Theorem Proving Approach. *IEEE Access* **2019**, *7*, 16550–16560. [CrossRef]

34. Qin, J.; Zhang, H.; Wang, S.; Geng, Z.; Chen, T. Acteve++: An Improved Android Application Automatic Tester Based on Acteve. *IEEE Access* **2019**, *7*, 31358–31363. [CrossRef]

35. Anand, S.; Naik, M.; Harrold, M.J.; Yang, H. Automated concolic testing of smartphone apps. In Proceedings of the 20th ACM SIGSOFT Symposium on the Foundations of Software Engineering (FSE-20), Cary, NC, USA, 11–16 November 2012; p. 59. [CrossRef]

36. Nisi, D.; Bianchi, A.; Fratantonio, Y. Exploring Syscall-Based Semantics Reconstruction of Android Applications. In Proceedings of the 22nd International Symposium on Research in Attacks, Intrusions and Defenses, Beijing, China, 23–25 September 2019; pp. 517–531.

37. Chin, E.; Felt, A.P.; Greenwood, K.; Wagner, D. Analyzing Inter-application Communication in Android. In Proceedings of the 9th International Conference on Mobile Systems, Applications, and Services, New York, NY, USA, June 2011; pp. 239–252. [CrossRef]

38. Drake, J.J.; Lanier, Z.; Mulliner, C.; Fora, P.O.; Ridley, S.A.; Wicherski, G. *Android Hacker's Handbook*; Wiley Publishing: Hoboken, NJ, USA, 2014.

39. Android Open Source Project (AOSP). Dalvik Bytecode. Available online: https://source.android.com/devices/tech/dalvik/dalvik-bytecode (accessed on 30 January 2020).

40. Oracle Corporation. Java Documentation on Thread. Available online: https://docs.oracle.com/javase/8/docs/api/java/lang/Thread.html (accessed on 2 October 2019).

41. Oracle Corporation. Java Documentation on Object. Available online: https://docs.oracle.com/javase/8/docs/api/java/lang/Object.html (accessed on 2 October 2019).

42. Göransson, A. *Efficient Android Threading: Asynchronous Processing Techniques for Android Applications*, 1st ed.; O'Reilly Media: Sebastopol, CA, USA, 2014; ISBN 978-1449364137.

43. Davi, L.; Dmitrienko, A.; Sadeghi, A.; Winandy, M. Privilege Escalation Attacks on Android. In Proceedings of the Information Security 13th International Conference, Boca Raton, FL, USA, 25–28 October 2010; pp. 346–360.

 *information*

Article

A Novel Low Processing Time System for Criminal Activities Detection Applied to Command and Control Citizen Security Centers

Julio Suarez-Paez [1,*], Mayra Salcedo-Gonzalez [1], Alfonso Climente [1], Manuel Esteve [1], Jon Ander Gómez [2], Carlos Enrique Palau [1] and Israel Pérez-Llopis [1]

[1] Distributed Real-time Systems Laboratory (SATRD), Universitat Politècnica de València, Camino de Vera s/n, 46022 Valencia, Spain; maysalgo@doctor.upv.es (M.S.-G.); alclial@upvnet.upv.es (A.C.), mesteve@dcom.upv.es (M.E.), cpalau@dcom.upv.es (C.E.P.), ispello0@upvnet.upv.es (I.P.-L.)

[2] Pattern Recognition and Human Language Technologies, Universitat Politècnica de València, Camino de Vera s/n, 46022 Valencia, Spain; jon@dsic.upv.es

* Correspondence: juliosuarez@ieee.org or jusuapae@doctor.upv.es

Received: 16 October 2019; Accepted: 20 November 2019; Published: 24 November 2019

Abstract: This paper shows a Novel Low Processing Time System focused on criminal activities detection based on real-time video analysis applied to Command and Control Citizen Security Centers. This system was applied to the detection and classification of criminal events in a real-time video surveillance subsystem in the Command and Control Citizen Security Center of the Colombian National Police. It was developed using a novel application of Deep Learning, specifically a Faster Region-Based Convolutional Network (R-CNN) for the detection of criminal activities treated as "objects" to be detected in real-time video. In order to maximize the system efficiency and reduce the processing time of each video frame, the pretrained CNN (Convolutional Neural Network) model AlexNet was used and the fine training was carried out with a dataset built for this project, formed by objects commonly used in criminal activities such as short firearms and bladed weapons. In addition, the system was trained for street theft detection. The system can generate alarms when detecting street theft, short firearms and bladed weapons, improving situational awareness and facilitating strategic decision making in the Command and Control Citizen Security Center of the Colombian National Police.

Keywords: Command and Control Citizen Security Center; Command and Control Information System (C2IS); crime detection; homeland security

---

1. Introduction

Colombia is a country with approximately 49 million inhabitants, 77% of which live in cities [1], and as in many Latin American countries, some Colombian cities suffer from insecurity. To face this situation and guarantee the country's sovereignty, the Colombian government has public security forces formed by the National Army, the National Navy and the Air Force, which have the responsibility to secure the borders of the country as well as ensure its sovereignty. Additionally, the Colombian National Police has the responsibility of security in the cities and of fighting against crime.

To ensure citizen security, the Colombian National Police has a force of 180,000 police officers, deployed across the national territory and several technological tools, such as Command and Control Information Systems (C2IS) [2,3] that centralize all the strategic information in real time, improving *situational awareness* [2,3] for making strategic decisions [3,4], such as the location of police officers and mobility of motorized units.

The C2IS centralizes the information in a physical place called the Command and Control Citizen Security Center (in Spanish: Centro de Comando y Control de Seguridad Ciudadana), where under a strict command line, the information is received by the C2IS operators and transmitted to the commanders of the National Police to make the most relevant operative decisions in the shortest time possible (Figure 1).

Figure 1. Command and Control Citizen Security Center, Colombian National Police.

The C2IS shows georeferenced information using a Geographic Information System (GIS) of several subsystems [5], such as crime cases reported by emergency calls, the position of the police officers in the streets and real-time video from the video surveillance system [6].

However, this technological system has a weakness in the Video Surveillance Subsystem because of the discrepancy between the number of security cameras in the Colombian cities and the system operators, which hinders the detection of criminal events. In other words, there are many more cameras than system operators can handle, meaning that the video information arrives at the Command and Control Citizen Security Center but it cannot be processed fast enough by the police commanders, and as such, they cannot take the necessary tactical decisions.

Bearing this in mind, this paper shows a Low Processing Time System focused on criminal activities detection based on real-time video analysis applied to a Command and Control Citizen Security Center. This system uses a novel method for detecting criminal actions, which applies an object detector based on Faster Region-Based Convolutional Network (R-CNN) as a detector of criminal actions. This innovative application of Faster R-CNN as a criminal action detector was achieved by training and adjusting the system for criminal activities detection using data extracted from the Command and Control Center of the Colombian National Police.

This novel method automates the detection of criminal events captured by the video surveillance subsystem, generating alarms that will be analyzed by the C2IS operators, improving situational awareness of the police commanders present at the Command and Control Citizen Security Center.

2. Related Work in Crime Events Video Detection

In computer vision, there are many techniques and applications which could be relevant for the operators of the C2IS of the National Police, for instance, the detection of pedestrians, the detection of trajectories, background and shadow removing [7], and facial biometrics.

There are already several approaches to detect crimes and violence in video analysis, as shown by [8–11]. However, the Colombian National Police does not implement any method for the specific case of the detection of criminal events. The available solutions are not applicable because most of the cameras of the video surveillance system installed in Colombian cities are mobile (*Pan–Tilt–Zoom Dome*), which makes it difficult to use conventional video analysis techniques focused on human action recognition because most of these methods are based on trajectory [12–15] or movement analysis [16–18] and camera movements interfere with these kinds of studies.

Owing to this, we decided to explore innovative techniques independent of the abrupt movement of video cameras, which perform a frame-by-frame analysis without independence between video frames.

Bearing this in mind, we discarded all the techniques based on trajectory detection and used prediction filters or metadata included in the video files, focusing on techniques that could take advantage of hardware's capabilities for parallel processing. As such, the criminal events detection system was developed using Deep Learning techniques.

Taking into account the technological developments of recent years, Deep Learning has become the most relevant technology for video analysis and has an advantage over the other technologies analyzed for this project: each video frame is analyzed and processed independently of all the others without temporary interdependence, which makes Deep Learning perfect for video analysis from mobile cameras such as those used in this project.

To choose the Deep Learning Models, we studied factors such as the processing time of each video frame, accuracy and model robustness. Therefore, several detection techniques were studied, such as R-CNN (Region-Based Convolutional Network) [19], YOLO (You Only Look Once) [20], Fast R-CNN (Fast Region-Based Convolutional Network) [21,22] and Faster R-CNN (Faster Region-Based Convolutional Network) [23,24] (Table 1). After analyzing the advantages and disadvantages of each technique, Faster R-CNN was chosen to implement the system for criminal events detection in the system for the C2IS of the National Colombian Police due to the fact that it has an average timeout that was 250 times faster than *R-CNN* and 25 times faster than Fast R-CNN [22,25,26]. Furthermore, in recent work, models based on two stages like Faster R-CNN have had better accuracy and stability than models based on regression like YOLO [27,28] and SSD, which is of great importance because in this work, a novel application focused in action detection was given to an object detector model.

Table 1. Deep Learning object detection models relative comparison.

Object Detector Model	Average Accuracy	Average Processing Time	Model Deployment Level (Number of Works Related)
R-CNN	High	High	Medium
Fast R-CNN	High	Medium	Medium
Faster R-CNN	Very High	Very Low	Very High
SSD	Very High	Very Low	Very High
YOLO	Very High	Very Low	Very High

Analyzing real-time video frame-by-frame is a task with a very high computational cost. This is considerable taking into account the sheer amount of video cameras surveillance systems available in Colombian cities. Therefore, it is necessary that each video frame has a low computational cost and processing time to secure a future large-scale implementation.

With this in mind, several previous studies have been studied where real-time video is analyzed with security applications. Among these studies, one stands out [29], in which the authors performed video analysis from a video surveillance system using the Caffe Framework [30] and Nvidia cuDNN [31] without using a supercomputer. Another study that demonstrated the high performance of Faster R-CNN for video analysis in real time is [32], in which the video was processed at a rate of 110 frames per second. Another interesting study is [33], in which the authors made a system based on Faster R-CNN for the real-time detection of evidence in crime scenes. One last study to highlight is [34], in which the authors created an augmented reality based on Faster R-CNN implementation using a gaming laptop.

Other authors have carried out related relevant research, such as [35], in which fire smoke was detected from video sources; [36], which showed a fire detection system based on artificial intelligence; [37], which detected terrorist actions on videos; [38,39], that showed novel applications to object detection; [40,41], that showed an excellent tracking applications; [42] in which a Real-Time video analysis was made from several sources with interesting results in object tracking; [43] which proposed a secure framework for IoT Systems Using Probabilistic Image Encryption; [44] which showed an Edge-Computing Video Analytics system deployed in Liverpool, Australia; [45] where GPUs and Deep Learning were used for traffic prediction; [46] where a video monitor and a radiation detector in nuclear accidents were shown; [47] where an Efficient IoT-based Sensor Big Data system was detailed.

In addition to these, recently, interesting applications of Faster R-CNN have also been published, for example in [48], a novel application of visual questions answering by parameter prediction using Faster R-CNN was presented, [49] showed a modification of Faster R-CNN for vehicles detection which improves detection performance, in [50], a face detection application was presented in low light conditions using two-step Faster R-CNN processing, first detecting bodies and then detecting faces, [51] showed an application to detect illicit objects such as fire weapons and knives, analyzing terahertz imaging using Faster R-CNN as an object detector and [52] showed a Faster R-CNN application for the detection of insulators in high-power electrical transmission networks.

As shown previously, Deep Learning includes a variety of techniques in computer vision, which are suitable for the development of this work.

3. Novel Low Computational Cost Method for Criminal Activities Detection Using One-Frame Processing Object Detector

In many cases, the detection and recognition of human actions (like criminal actions) is done by analysis of movement [16–18,53,54] or trajectories [12–15], which implies the processing of several video frames. Nevertheless, when the video camera is mobile, it is very difficult to carry out the trajectory or movement analysis because camera movements may introduce noise to the trajectories or movements to be analyzed. In addition, in a Smart City application, the number of cameras could be hundreds or thousands, so motion or trajectories analysis involves processing several video frames for each detection, which would multiply the computational cost of a possible solution. It is necessary to analyze mobile cameras with the minimum computational cost possible because, in the Command and Control Citizen Security Center, thousands of cameras are pan–tilt–zoom domes and this makes it very difficult to perform a motion or trajectory analysis to detect criminal activities. On the other hand, since there are thousands of cameras, the computational cost becomes an extreme relevant factor.

For this reason, hours of video of criminal activities were studied and it was noted that all criminal activities have a characteristic gesture, such as threatening someone; therefore, we set out to analyze this characteristic gesture as an "object" so that it could be detected using techniques that are independent of camera movements and process only one video frame.

With this in mind, we propose a novel system called "Video Detection and Classification System (VD&CS)" in which Faster R-CNN is used in a hybrid way to detect objects used in criminal actions and criminal characteristic gestures treated as "objects". Considering that criminal actions always have fixed gestures such as threatening the victim, it is possible to consider that this criminal action can be understood by the system as an "object". This novel application has the potential to reduce the computational cost because only one video frame will be processed, compared to other action detection methods that must analyze several video frames [12–18,53,54]. With this novel method in mind, we proceeded with the system design and training.

3.1. Video Detection and Classification System (VD&CS)

The system proposed is based on a Faster Region-Based Convolutional Network (Faster R-CNN), involves two main parts: a region proposal network (RPN) and a Fast R-CNN [23] and it was developed using Matlab.

3.1.1. Region Proposal Network

The RPN is composed of a classifier and a regressor, and its aim is to predict whether, in a certain image region, a detectable object will exist or will be part of the background, as is shown in [23].

Regions of interest comprise short firearms, bladed weapons and street thefts, which are criminal actions but will be treated as objects in the training process.

In this case, the pre-trained CNN model AlexNet [55] was used as the core of the RPN. This CNN model is made up of Convolution layers, ReLU, Cross Channel Normalization layers, Max Pool layers, Fully Connected layers and Softmax layers, as shown in Figure 2.

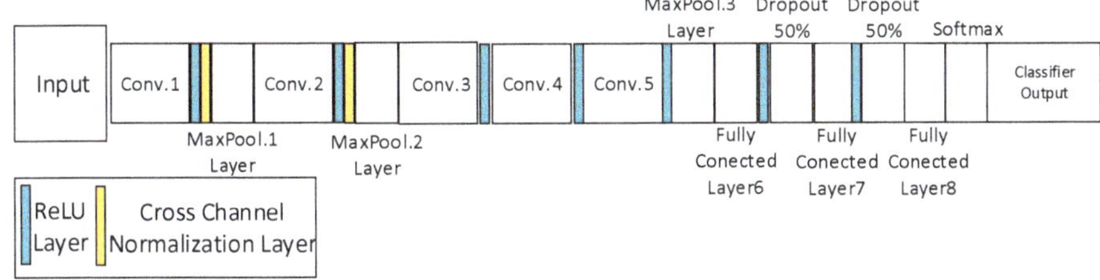

Figure 2. AlexNet Convolutional Neural Network Layers [55].

Figure 3 shows AlexNet used as RPN core. It has less layers than models like VGG16 [56], VGG19 [56], GoogleNet [57] or ResNet [58]. Hence, AlexNet has a lower computational cost and requires less processing time per video frame [22] (further implementation details are provided in Section 5).

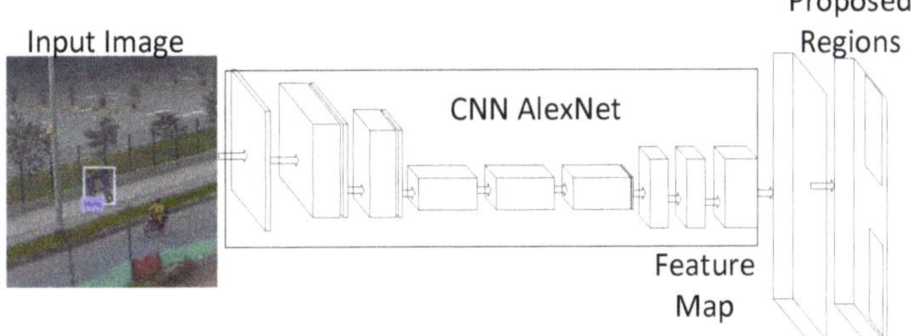

Figure 3. Video Detection and Classification System (VD&CS): Region Proposal Network (RPN).

3.1.2. Fast Region-Based Convolutional Network

Fast R-CNN acts as a detector that uses the region proposals made by the RPN and also uses AlexNet (Figure 2) as the CNN of the core model to detect regions of interest for the system, which are short firearms, bladed weapons and street thefts (Figure 4).

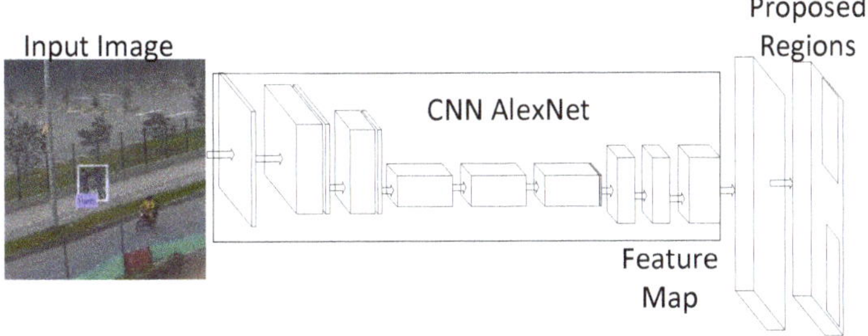

Figure 4. VD&CS: fast R-CNN.

3.1.3. Faster Region-Based Convolutional Network

Finally, RPN and Fast R-CNN are joined, forming a Faster R-CNN system (Figure 5) with the capacity of real-time video processing using AlexNet (Figure 2) as core.

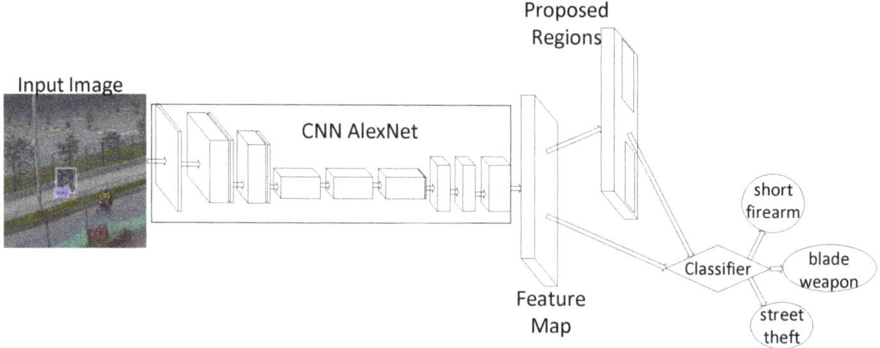

Figure 5. VD&CS: faster R-CNN.

3.2. VD&CS: Training Process

The system proposed based on Faster R-CNN, was trained using Matlab in a four-stage process, as outlined below.

3.2.1. Train RPN Initialized with AlexNet Using a New Dataset

At this stage, AlexNet, shown in Figure 2, is retrained inside the RPN, using transfer learning with a new dataset of 1124 images specially created to train the VD&CS (Figure 6). This dataset was created by manually analyzing several hours of video taken from the Command and Control Citizen Security Center and finding criminal actions to extract. The dataset has three classes of interest: short firearm, bladed weapons and street theft (action as object), and its bound boxes were manually marked for each image.

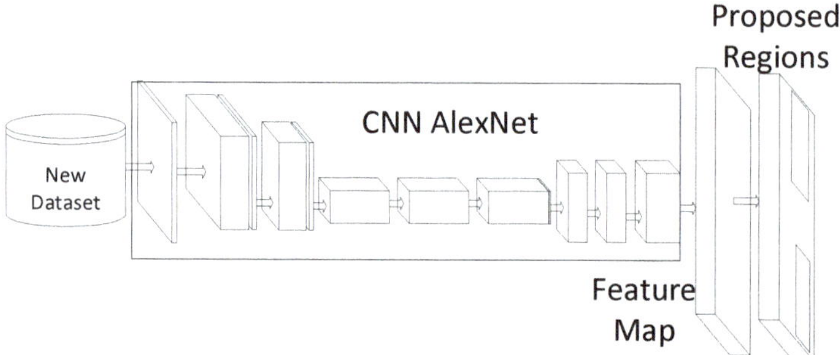

Figure 6. First stage: RPN training.

To improve the system performance, the training process data argumentation methods were used and as a result of the training procedure, in this stage, we obtained a feature map of the three classes mentioned above, from which the RPN is able to make proposals of possible regions of interest.

3.2.2. Train Fast R-CNN as a Detector Initialized with AlexNet Using the Region Proposal Extracted from the First Stage

In the second stage, a Fast R-CNN detector was trained using the initialized AlexNet as a starting point (Figure 7). The region proposals obtained by the RPN in the first stage were used as input to the Fast R-CNN to detect the three classes of interest.

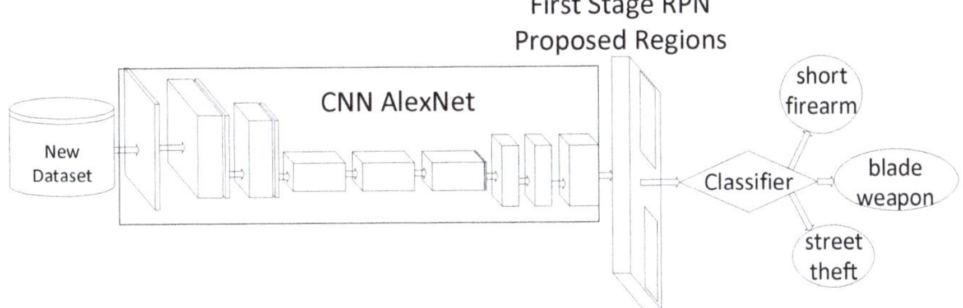

Figure 7. Second stage: Fast R-CNN training.

3.2.3. RPN Fine Training Using Weights Obtained with Fast R-CNN Trained in the Second Stage

With the objective of increasing the RPN success rate, in the third stage, fine training of the RPN that was trained in the first stage is carried out (Figure 8). In this case, weights obtained from the training procedure of the Fast R-CNN during the second stage were used as initial values.

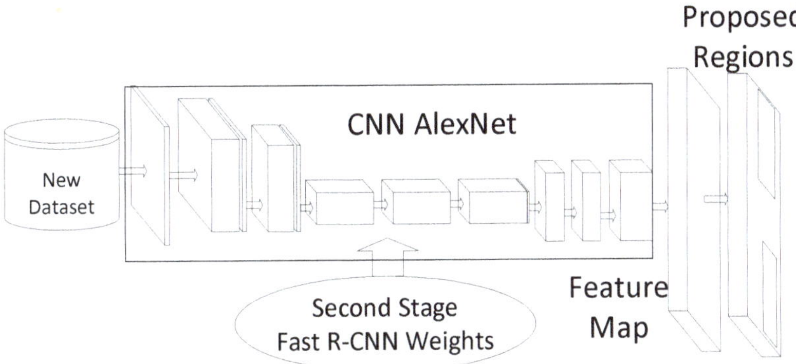

Figure 8. Third stage: RPN fine training.

3.2.4. Fast R-CNN Fine Training Using Updated RPN

To improve the accuracy of the Fast R-CNN trained in the second stage, in this last stage, fine training was carried out using the results of the third stage, as shown in Figure 9.

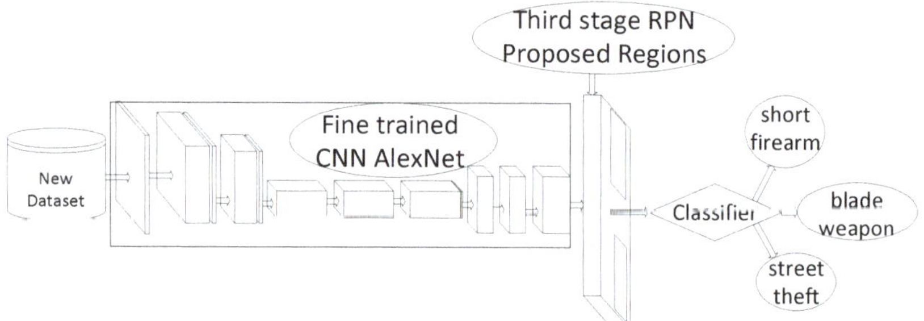

Figure 9. Fourth stage: Fast R-CNN fine training.

Finally, Figure 10 shows a system capable at generating alarms detecting short weapons, blade weapons and street theft by analyzing just one video frame, which would reduce the computational cost compared to models based on analysis of movement or trajectories.

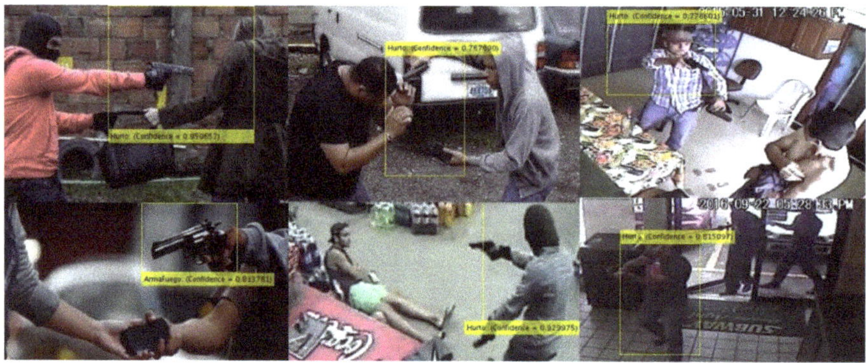

Figure 10. VD&CS: criminal activities detection.

3.3. VD&CS: Testing

Once VD&CS was trained, its image processing time and accuracy were measured in order to evaluate its applicability to real scenarios of real-time video analysis. Two series of 500 images that were not used for training were used for testing using the same Hardware: MSI GT62VR-7RE with an Intel Core I7 7700HQ, 16 GB of DDR4 RAM, with a GPU NVIDIA GeForce GTX 1070 with 8 GB DDR5 VRAM). Table 2 shows the obtained results. The used performance indicators were the average processing time per frame, accuracy, undetected event rate, false positive rate and frame rate per second (FPS).

Table 2. VD&CS 500 image tests.

Item Tested	Results Test 1	Results Test 2
Crime Event Detections	355	367
Failures	145	133
Undetected	87	80
False positive	58	53
Average processing time	0.03 s	0.03 s
FPS (Frames per second)	33 FPS	33 FPS
Undetected event rate	17.4%	16%
False positive rate	11.6%	10.6%
Accuracy	71%	73.4%

According to previous results, the Confusion Matrix (Table 3) shows that VD&CS is useful for detecting criminal events in real-time video; its accuracy is within the parameters expected of a Faster R-CNN [23], taking into account that criminal actions were handled as objects within VD&CS, it confirms that VD&CS can be used for Criminal Activities Detection Applied to Command and Control Citizen Security Centers.

Table 3. VD&CS confusion matrix.

	Predictions	
Observations	49.6% (True Positive)	11.6% (False Positive)
	17.4% (False Negative)	21.4% (True Negative)

3.3.1. Real-Time Video Testing

Training and tests with real-time video were performed on a laptop MSI GT62VR-7RE with an Intel Core I7 7700HQ, 16 GB of DDR4 RAM, with a GPU NVIDIA GeForce GTX 1070 with 8 GB DDR5 VRAM.

Real-time video testing consisted of two main video sources; the first source contained pre-recorded videos obtained from the Colombian National Police video surveillance system and the second source was a set of videos captured in real time by a laptop camera.

In these two scenarios, excellent results were obtained with respect to the processing time of each image, ranging from 0.03 to 0.05 s. This allows real-time video processing at a rate of 20 to 33 FPS, which is adequate considering the video sources of the C2IS of Colombian National Police.

Regarding the system accuracy, we checked that it is free of overtraining as the tests done on the system were performed with images not used in the training process and their results were confirmed in the confusion matrix and the system accuracy it is within the range expected for a Faster R-CNN; however, the system is designed to be used in public safety applications, so it always requires human monitoring because the detections depend on the lighting conditions and the distance of the cameras to the object, in addition to the success rate of the Faster R-CNN; additionally, in previous studies [22], authors evaluated other CNN models of a greater depth by choosing AlexNet for its performance and simplicity.

However, it achieves excellent results in terms of triggering alarms when it detects criminal events, improving situational awareness in the Command and Control Citizen Security Center of Colombian National Police.

3.3.2. Computational Cost Comparation

As previously stated, several detection and recognition of human actions techniques consist of movement or trajectories analysis. These techniques must analyze several video frames to be able to recognize actions, for example, in [59–61], sets of six to eight images are analyzed to identify actions.

In order to have computational cost low enough to be deployed in thousands of cameras, VD&CS just processes one video frame to detect criminal actions, which achieves a low computational cost that could be deployed in embedded systems or in cloud architecture, reducing high deployment costs.

To analyze the computational cost, different CNN models in the VD&CS core were compared with another action detection technique proposed in [59]. The results are shown in Table 4.

Table 4. VD&CS computational cost comparison.

Model	Average Processing Time	GPU	GPU Performance (Float 32)	Resolution (Pixels)
VD&CS (AlexNET)	0.03 s	Nvidia GTX 1070 MXM	6.738 TFLOPS	704×544
VD&CS (VGG-16)	0.23 s	Nvidia GTX 1070 MXM	6.738 TFLOPS	704×544
VD&CS (VGG-19)	0.28 s	Nvidia GTX 1070 MXM	6.738 TFLOPS	704×544
T-CNN	0.9 s	Nvidia GTX Titan X	6.691 TFLOPS	300×400

Therefore, assuming that GPUs have an equivalent performance and scaling the resolution of the video frames used in the tests, we consider deployments in cities like Bogotá where there are about 2880 Pan–Tilt–Zoom cameras (as of June 2019).

First, we analyzed computational cost measured TeraFlops and depict the variation of computational costs for processing of 2880 cameras (Figure 11).

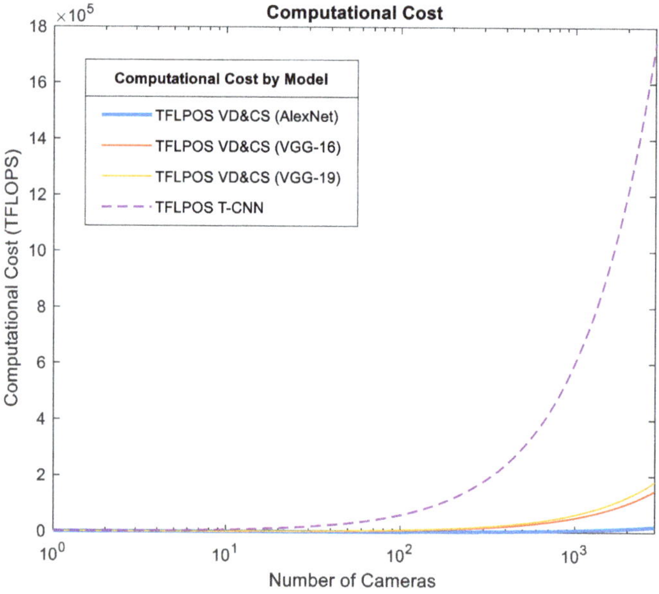

Figure 11. VD&CS low processing time system: computational cost comparison.

We also analyzed the Hardware cost and power consumption, assuming a deployment using Nvidia embedded systems [62] (Figures 12 and 13).

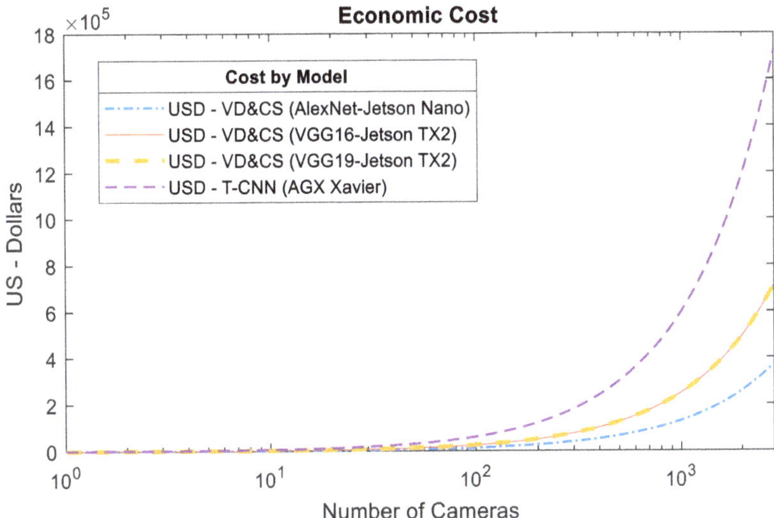

Figure 12. VD&CS Low Processing Time System: Economical Cost Comparation.

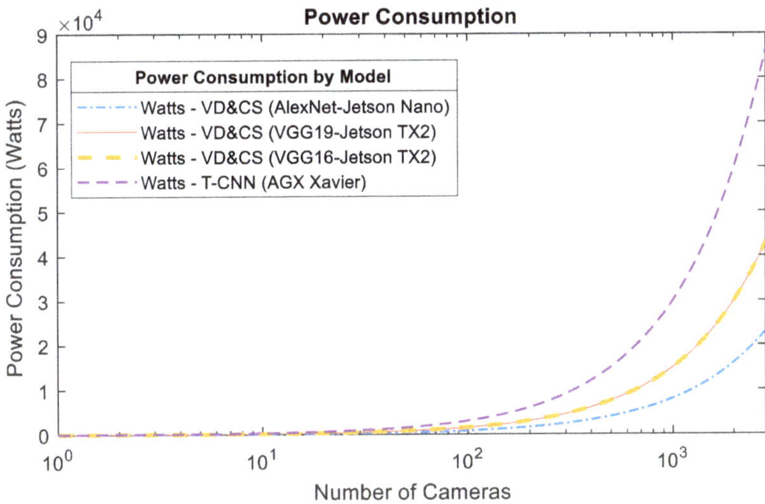

Figure 13. VD&CS Low Processing Time System: Power Consumption Comparation.

As Figures 11–13 show, having thousands of video sources in a Low Processing Time System, the computational cost is a factor of extreme relevance, since the economic and energy costs could make the implementation not feasible, and for this reason VD&CS proves be appropriate in a Low Cost System.

3.4. VD&CS: Final System

Once the process of training and testing are completed, we propose the system shown in Figure 14 to be applied in a larger city architecture.

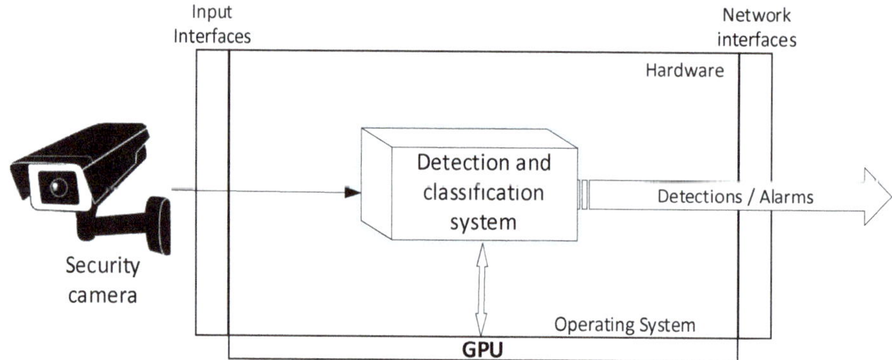

Figure 14. VD&CS: System.

In this approach, the VD&CS runs in an environment independent of the operating system because it can be implemented using any framework or library that supports Faster R-CNN, such as Caffe [30], cuDNN [31], TensorFlow [63], TensorRT [64], Nvidia DeepStream SDK [65], which uses real-time video coming from the security cameras and uses GPU computational power to run. Finally, the VD&CS uses network interfaces to send the generated alarms to the Command and Control Citizen Security Center.

This system is expected to be applied in different scenarios based on cloud architectures or embedded systems compatible with IoT (Internet of Things) solutions [66,67].

4. Low Processing Time System Applied to Colombian National Police Command and Control Citizen Security Center

To propose a Low Processing Time System to detect criminal activities based on a real-time video analysis applied to National Police of Command and Control Citizen Security Center, we must consider the Colombian Police Command and Control objectives, as detailed below:

Situational awareness: Police commanders must know in detail and real-time the situation of citizen security in the field, supported by technological tools to make the best tactical decisions and guarantee the success of police operations that ensure citizen security.

Situation understanding: Improving situational awareness by improving crime detection, allows police commanders to gain a better understanding of the situation, helping to detect more complex behaviors of criminal gangs.

Decision making-improvement: Decisions made in the Command and Control Citizen Security Center can be life or death because many criminal acts involve firearms and violent acts; therefore, the proposed system will improve decision making because it will provide real-time information to commanders, improving the effectiveness of police operations.

Agility and efficiency improvement: As mentioned above, decisions made by the police can mean life or death. Therefore, the improvement offered by the proposed prototype to the agility and efficiency of police operations relies on information that is unknown by commanders, impeding the deployment of police officers in critical situations.

4.1. Decentralized Low Processing Time System for Criminal Activities Detection based on Real-time Video Analysis Applied to the Colombian National Police Command and Control Citizen Security Center

The Command and Control Citizen Security Center is formed of subsystems such as the emergency call attention system (123), Police Cases Monitoring and Control Information System (SECAD), Video Surveillance Subsystem and the crisis and command room. Command and Control Citizen Security is supported by telecom networks that can be owned by the National Police or belong to the local ISP (Internet Service Provider).

These subsystems have different types of operators which are in charge of specific tasks such as monitoring the citizen security video (Operators Video Surveillance system), answering emergency calls (123 Operators) and assigning and monitoring field cop to police cases (Dispatchers).

Another important part of the Command and Control Citizen Security Center is the crisis and command room, in which the police commanders make strategic decisions according to their situational awareness and situation understanding [2].

In this decentralized system, the VD&CS will be implemented in embedded systems with GPU capability such as Nvidia Jetson [62] or AMD Embedded Radeon™ [68]. Then, it will be installed in each citizen video surveillance camera, detecting criminal activities locally (Figure 15).

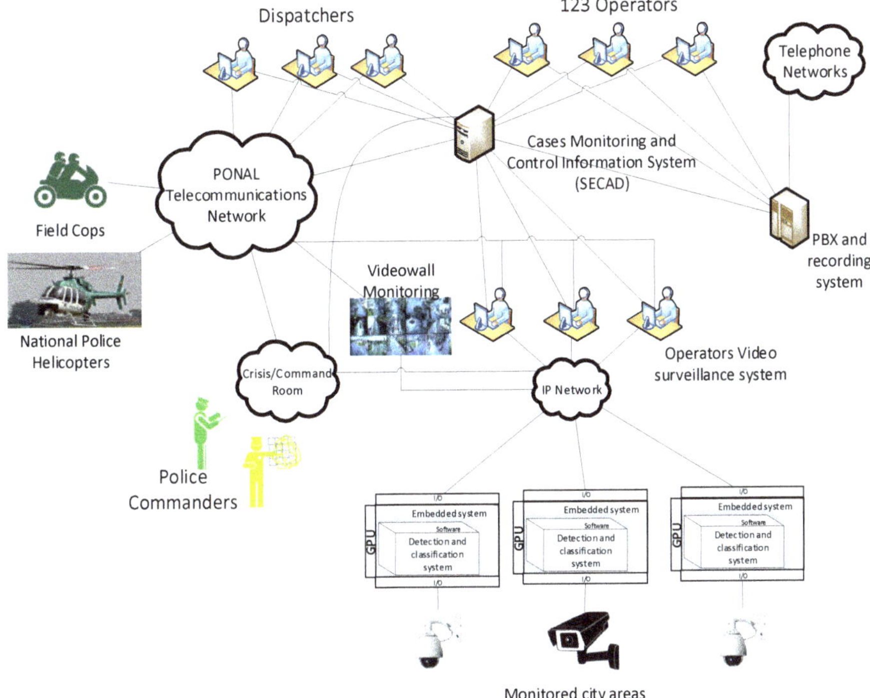

Figure 15. Decentralized Low Processing Time System for criminal activities detection.

After each detection, alarms will be generated and will be sent by a network to the Video Surveillance Subsystem where operators can take actions to prevent and respond to criminal actions.

4.2. Centralized Low Processing Time System to Criminal Activities Detection Based on Real-Time Video Analysis Applied to Colombian National Police Command and Control Citizen Security Center

In contrast to the previously decentralized system shown before, in this case, the video will be processed in a centralized infrastructure with high computational power and GPU capabilities (Figure 16).

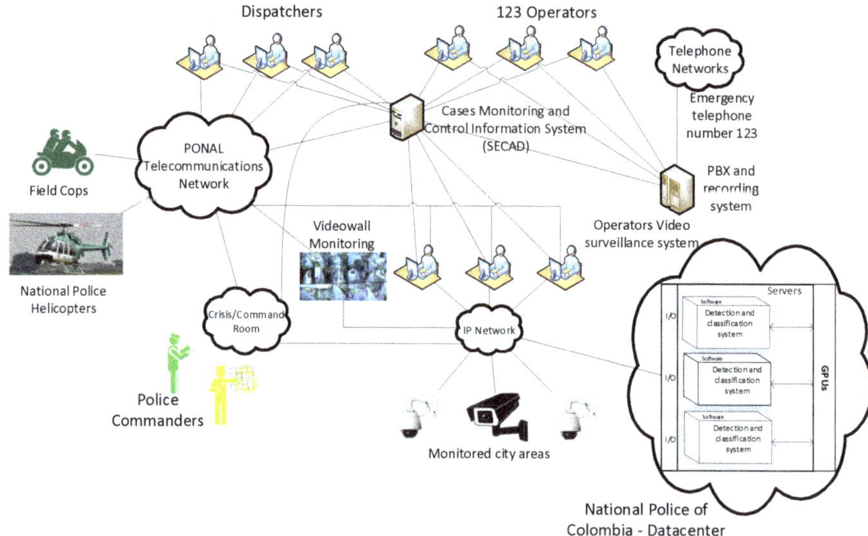

Figure 16. Centralized Low Processing Time System to criminal activities detection.

The datacenter runs the VD&CS individually for each video signal coming from each of the city video surveillance cameras, generating alarms when criminal activities are detected, and sends it back to the Video Surveillance Subsystem through the network, where operators can take actions to prevent and respond to criminal actions.

5. Possible Implementation and Limitations

Considering that the development of VD&CS was performed on a laptop using Matlab and Windows 10 and that an image processing rate of 20 to 30 frames per second was obtained, it is feasible to migrate the VD&CS to an environment with greater efficiency using the libraries optimized for Deep Learning, such as cuDNN [31], TensorFlow [63], TensorRT [64], Nvidia DeepStream SDK [65], further reducing the computational cost.

With this reduction in the computational cost, it would be possible to implement VD&CS in embedded systems, such as the Nvidia Jetson [62] and optimize the implementation using Nvidia DeepStream [65], to be installed directly in citizen security cameras and subsequently generate alerts upon the occurrence of criminal events which would be reported to the Command and Control Citizen Security Center of the Colombian National Police, like in the decentralized Low Processing Time System.

Currently, in June 2019 in Bogotá D.C., there are about 2880 Pan–Tilt–Zoom cameras that are monitored in the Citizen Security Control Center, and these domes generate around 22.4 Gbps of real-time video traffic. Given that currently, in Colombia, there is no cloud provider that has datacenters in the country, it would not be applicable to use cloud solutions with datacenters in the United States or Brazil because the international channel cost would be very high; therefore, in June 2019, the best solution is to use embedded systems, at least until a cloud provider provides a datacenter with GPU capability in Colombia.

The VD&CS limitations must be considered in future implementations because, like all systems based on Deep Learning, it is not 100% reliable and its precision is linked to critical factors such as lighting and partial obstructions, meaning that human supervision is necessary.

However, the implementation of this Low Processing Time System in a large-scale environment depends on the budget availability of the Government of Colombia.

6. Discussion and Future Application

VD&CS have proven to be effective in a hybrid operation as an object detector and the treatment of criminal actions as objects. If the characteristic gestures are identified in certain actions, it should be possible to use object detectors based on Deep Learning in various applications such as the detection of suspicious activities, fights, riots and more.

As shown above, several recent applications of Faster R-CNN have shown great performance as object detector [48–52], however, this work demonstrated that applying object detection techniques based on Deep Learning like Faster R-CNN in actions detection could be an alternative to action recognition based on analysis of trajectories or movements and could be applied more easily in highly mobile video environments, such as military operations, transportation, citizen security, and national security to name only a few, nevertheless, human supervision is always required, because after a while, the quantity of False Negatives and False Positive could drastically reduce the system effectiveness, which is very serious in safety applications.

In future research, we could identify human actions that could be recognized using object detectors based on Deep Learning.

These actions should have characteristic gestures like in the case of criminal activities, which always have recognizable gestures such as threatening the victim.

Although the system's accuracy is around 70%, this percentage can be considered acceptable because the system is tolerant to the sudden movements of the Pan–Tilt–Zoom cameras of the Colombian National Police. It also shows that it is possible to use an object detector to detect criminal actions and in future applications, the system's accuracy could be improved.

Further future research work consists of maximizing the recognition of human actions using an objects classifier, minimizing system failures. This can be achieved by building more complete datasets and experimenting with diverse Deep Learning techniques such as YOLO, and several CNN models such as ResNet, GoogleNet.

7. Conclusions

By applying the secure city architectures in command and control systems, situational awareness and situation understanding of police commanders will improve, as well as their agility and efficiency in decision making, thus improving the effectiveness of police operations and directly increasing citizen security.

During the development of the VD&CS, it has been proven that it is possible to improve situational awareness in the Command and Control Citizen Security Center of the Colombian National Police, triggering alarms of criminal events captured by the video surveillance system.

Reducing the computational cost for using Deep Learning or any other technique in citizen security applications is fundamental for achieving real-time performance and feasible implementation costs, especially given the amount of information generated by surveillance systems. The processing time is vital to achieve a real improvement of situational awareness.

The Low Processing Time System to Criminal Activities Detection Applied to a Command and Control Citizen Security Center could be deployed in Colombia because the VD&CS showed that it is possible to detect criminal actions using a Deep Learning Object Detector as long as the system is trained to detect actions (these actions must have characteristic gestures such as threatening the victim). Deep Learning can be a powerful tool in citizen security systems because it can automate the detection of situations of interest which can escape from system operator view in Command and Control Information System of a security agency such as the Colombian National Police.

Author Contributions: Conceptualization, J.S.-P., M.S.-G. and A.C.; methodology, M.E., J.A.G. and C.E.P.; software, J.S.-P. and, M.S.-G.; validation, A.C., and I.P.-L.; formal analysis, J.S.-P., M.S.-G. and A.C.; investigation, J.S.-P., M.S.-G. and A.C.; writing—original draft preparation, J.S.-P., M.S.-G. and A.C.; writing—review and editing, J.S.-P., M.S.-G. and A.C.; visualization, J.S.-P., M.S.-G. and A.C.; supervision, M.E., C.E.P. and J.A.G.; project administration, I.P.-L.; funding acquisition, M.E., C.E.P. and I.P.-L.

Funding: This work was co-funded by the European Commission as part of H2020 call SEC-12-FCT-2016-Subtopic3 under the project VICTORIA (No. 740754). This publication reflects the views only of the authors and the Commission cannot be held responsible for any use which may be made of the information contained therein.

Acknowledgments: The authors thank Colombian National Police and its Office of Telematics for their support on the development of this project.

Conflicts of Interest: The authors declare no conflict of interest.

References

1. World Bank United Nations. *Perspectives of Global Urbanization*; Command and Control and Cyber Research Portal (CCRP): Washington, DC, USA, 2019.
2. Alberts, D.S.; Hayes, R.E. *Understanding Command and Control the Future of Command and Control*; Command and Control and Cyber Research Portal (CCRP): Washington, DC, USA, 2006.
3. Esteve, M.; Perez-Llopis, I.; Hernandez-Blanco, L.E.; Palau, C.E.; Carvajal, F. SIMACOP: Small Units Management C4ISR System. In Proceedings of the IEEE International Conference Multimedia and Expo, Beijing, China, 2–5 July 2007; pp. 1163–1166.
4. Wang, L.; Rodriguez, R.M.; Wang, Y.-M. A dynamic multi-attribute group emergency decision making method considering experts' hesitation. *Int. J. Comput. Intell. Syst.* **2017**, *11*, 163. [CrossRef]
5. Esteve, M.; Perez-Llopis, I.; Palau, C.E. Friendly force tracking COTS solution. *IEEE Aerosp. Electron. Syst. Mag.* **2013**, *28*, 14–21. [CrossRef]
6. Esteve, M.; Pérez-Llopis, I.; Hernández-Blanco, L.; Martinez-Nohales, J.; Palau, C.E. Video sensors integration in a C2I system. In Proceedings of the IEEE Military Communications Conference MILCOM, Boston, MA, USA, 18–21 October 2009; pp. 1–7.
7. Spagnolo, P.; D'Orazio, T.; Leo, M.; Distante, A. Advances in background updating and shadow removing for motion detection algorithms. In *Lecture Notes in Computer Science (Including Subseries Lecture Notes in Artificial Intelligence and Lecture Notes in Bioinformatics), Proceedings of the 11th International Conference on Computer Analysis of Images and Patterns, CAIP 2005, Versailles, France, 5–8 September 2005*; Springer: Versailles, France, 2005; Volume 3691, pp. 398–406.
8. Nieto, M.; Varona, L.; Senderos, O.; Leskovsky, P.; Garcia, J. Real-time video analytics for petty crime detection. In Proceedings of the 7th International Conference on Imaging for Crime Detection and Prevention (ICDP 2016), Madrid, Spain, 23–25 November 2017; pp. 23–26.
9. Senst, T.; Eiselein, V.; Kuhn, A.; Sikora, T. Crowd Violence Detection Using Global Motion-Compensated Lagrangian Features and Scale-Sensitive Video-Level Representation. *IEEE Trans. Inf. Forensics Secur.* **2017**, *12*, 2945–2956. [CrossRef]
10. Machaca Arceda, V.; Gutierrez, J.C.; Fernandez Fabian, K. Real Time Violence Detection in Video. In Proceedings of the International Conference on Pattern Recognition Systems (ICPRS-16), Talca, Chile, 20–22 April 2016; pp. 6–7.
11. Bilinski, P.; Bremond, F. Human violence recognition and detection in surveillance videos. In Proceedings of the 13th IEEE International Conference on Advanced Video and Signal Based Surveillance, AVSS 2016, Colorado Springs, CO, USA, 23–26 August 2016; pp. 30–36.
12. Xue, F.; Ji, H.; Zhang, W.; Cao, Y. Action Recognition Based on Dense Trajectories and Human Detection. In Proceedings of the IEEE International Conference on Automation, Electronics and Electrical Engineering (AUTEEE), Shenyang, China, 16–18 November 2018; pp. 340–343.
13. Shi, Y.; Tian, Y.; Wang, Y.; Huang, T. Sequential Deep Trajectory Descriptor for Action Recognition with Three-Stream CNN. *IEEE Trans. Multimed.* **2017**, *19*, 1510–1520. [CrossRef]
14. Dasari, R.; Chen, C.W. MPEG CDVS Feature Trajectories for Action Recognition in Videos. In Proceedings of the IEEE 1th International Conference on Multimedia Information Processing and Retrieval, Miami, FL, USA, 10–12 April 2018; pp. 301–304.
15. Arunnehru, J.; Chamundeeswari, G.; Bharathi, S.P. Human Action Recognition using 3D Convolutional Neural Networks with 3D Motion Cuboids in Surveillance Videos. *Procedia Comput. Sci.* **2018**, *133*, 471–477. [CrossRef]

16. Kamel, A.; Sheng, B.; Yang, P.; Li, P.; Shen, R.; Feng, D.D. Deep Convolutional Neural Networks for Human Action Recognition Using Depth Maps and Postures. *IEEE Trans. Syst. Man Cybern. Syst.* **2018**, *49*, 1–14. [CrossRef]

17. Ren, J.; Reyes, N.H.; Barczak, A.L.C.; Scogings, C.; Liu, M. Towards 3D human action recognition using a distilled CNN model. In Proceedings of the IEEE 3rd International Conference Signal and Image Processing (ICSIP), Shenzhen, China, 13–15 July 2018; pp. 7–12.

18. Zhang, B.; Wang, L.; Wang, Z.; Qiao, Y.; Wang, H. Real-Time Action Recognition with Deeply Transferred Motion Vector CNNs. *IEEE Trans. Image Process.* **2018**, *27*, 2326–2339. [CrossRef]

19. Girshick, R.; Donahue, J.; Darrell, T.; Malik, J. Region-Based Convolutional Networks for Accurate Object Detection and Segmentation. *IEEE Trans. Pattern Anal. Mach. Intell.* **2016**, *38*, 142–158. [CrossRef]

20. Redmon, J.; Divvala, S.; Girshick, R.; Farhadi, A. You only look once: Unified, real-time object detection. In Proceedings of the IEEE Computer Society Conference on Computer Vision and Pattern Recognition, Las Vegas, NV, USA, 26 June–1 July 2016; pp. 779–788.

21. Girshick, R. Fast R-CNN. In Proceedings of the IEEE International Conference on Computer Vision (ICCV), Washington, DC, USA, 3–7 December 2015; pp. 1440–1448.

22. Suarez-Paez, J.; Salcedo-Gonzalez, M.; Esteve, M.; Gómez, J.A.; Palau, C.; Pérez-Llopis, I. Reduced computational cost prototype for street theft detection based on depth decrement in Convolutional Neural Network. Application to Command and Control Information Systems (C2IS) in the National Police of Colombia. *Int. J. Comput. Intell. Syst.* **2018**, *12*, 123. [CrossRef]

23. Ren, S.; He, K.; Girshick, R.; Sun, J. Faster R-CNN: Towards Real-Time Object Detection with Region Proposal Networks. *IEEE Trans. Pattern Anal. Mach. Intell.* **2017**, *39*, 1137–1149. [CrossRef]

24. Hao, S.; Wang, P.; Hu, Y. Haze image recognition based on brightness optimization feedback and color correction. *Information* **2019**, *10*, 81. [CrossRef]

25. Jiang, H.; Learned-Miller, E. Face Detection with the Faster R-CNN. In Proceedings of the 12th IEEE International Conference on Automatic Face and Gesture Recognition, FG 2017—1st International Workshop on Adaptive Shot Learning for Gesture Understanding and Production, ASL4GUP 2017, Washington, DC, USA, 3 June 2017; pp. 650–657.

26. Peng, M.; Wang, C.; Chen, T.; Liu, G. NIRFaceNet: A convolutional neural network for near-infrared face identification. *Information* **2016**, *7*, 61. [CrossRef]

27. Wu, S.; Zhang, L. Using Popular Object Detection Methods for Real Time Forest Fire Detection. In Proceedings of the 11th International Symposium on Computational Intelligence and Design, ISCID, Hangzhou, China, 8–9 December 2018; pp. 280–284.

28. Chen, J.; Miao, X.; Jiang, H.; Chen, J.; Liu, X. Identification of autonomous landing sign for unmanned aerial vehicle based on faster regions with convolutional neural network. In Proceedings of the Chinese Automation Congress, CAC, Jinan, China, 20–22 October 2017; pp. 2109–2114.

29. Xu, W.; He, J.; Zhang, H.L.; Mao, B.; Cao, J. Real-time target detection and recognition with deep convolutional networks for intelligent visual surveillance. In Proceedings of the 9th International Conference on Utility and Cloud Computing—UCC '16, New York, NY, USA, 23–26 February 2016; pp. 321–326.

30. Jia, Y.; Shelhamer, E.; Donahue, J.; Karayev, S.; Long, J.; Girshick, R.; Guadarrama, S.; Darrell, T. Caffe: Convolutional architecture for fast feature embedding. In Proceedings of the ACM International Conference on Multimedia—MM '14, New York, NY, USA, 18–19 June 2014; pp. 675–678.

31. Nvidia Corporation. NVIDIA CUDA®Deep Neural Network library (cuDNN). Available online: https://developer.nvidia.com/cuda-downloads (accessed on 24 November 2019).

32. Song, D.; Qiao, Y.; Corbetta, A. Depth driven people counting using deep region proposal network. In Proceedings of the IEEE International Conference on Information and Automation, ICIA 2017, Macau SAR, China, 18–20 July 2017; pp. 416–421.

33. Saikia, S.; Fidalgo, E.; Alegre, E.; Fernández-Robles, L. Object Detection for Crime Scene Evidence Analysis Using Deep Learning. In *Lecture Notes in Computer Science (Including Subseries Lecture Notes in Artificial Intelligence and Lecture Notes in Bioinformatics), Proceedings of the International Conference on Mobile and Wireless Technology, ICMWT 2017, Kuala Lumpur, Malaysia, 26–29 June 2017*; Springer: Cham, Switzerland, 2017; pp. 14–24.

34. Sutanto, R.E.; Pribadi, L.; Lee, S. 3D integral imaging based augmented reality with deep learning implemented by faster R-CNN. In Proceedings of the Lecture Notes in Electrical Engineering, Proceedings of the

International Conference on Mobile and Wireless Technology, ICMWT 2017, Kuala Lumpur, Malaysia, 26–29 June 2017; Springer: Singapore, 2018; pp. 241–247.

35. Wu, X.; Lu, X.; Leung, H. A video based fire smoke detection using robust AdaBoost. *Sensors* **2018**, *18*, 3780. [CrossRef] [PubMed]

36. Park, J.H.; Lee, S.; Yun, S.; Kim, H.; Kim, W.-T.; Park, J.H.; Lee, S.; Yun, S.; Kim, H.; Kim, W.-T. Dependable Fire Detection System with Multifunctional Artificial Intelligence Framework. *Sensors* **2019**, *19*, 2025. [CrossRef]

37. García-Retuerta, D.; Bartolomé, Á.; Chamoso, P.; Corchado, J.M. Counter-Terrorism Video Analysis Using Hash-Based Algorithms. *Algorithms* **2019**, *12*, 110. [CrossRef]

38. Zhao, B.; Zhao, B.; Tang, L.; Han, Y.; Wang, W. Deep spatial-temporal joint feature representation for video object detection. *Sensors* **2018**, *18*, 774. [CrossRef]

39. He, Z.; He, H. Unsupervised Multi-Object Detection for Video Surveillance Using Memory-Based Recurrent Attention Networks. *Symmetry* **2018**, *10*, 375. [CrossRef]

40. Zhang, H.; Zhang, Z.; Zhang, L.; Yang, Y.; Kang, Q.; Sun, D.; Zhang, H.; Zhang, Z.; Zhang, L.; Yang, Y.; et al. Object Tracking for a Smart City Using IoT and Edge Computing. *Sensors* **2019**, *19*, 1987. [CrossRef]

41. Mazzeo, P.L.; Giove, L.; Moramarco, G.M.; Spagnolo, P.; Leo, M. HSV and RGB color histograms comparing for objects tracking among non overlapping FOVs, using CBTF. In Proceedings of the 8th IEEE International Conference on Advanced Video and Signal Based Surveillance, AVSS 2011, Washington, DC, USA, 30 August–2 September 2011; pp. 498–503.

42. Leo, M.; Mazzeo, P.L.; Mosca, N.; D'Orazio, T.; Spagnolo, P.; Distante, A. Real-time multiview analysis of soccer matches for understanding interactions between ball and players. In Proceedings of the International Conference on Content-based Image and Video Retrieval, Niagara Falls, ON, Canada, 7–9 July 2008; pp. 525–534.

43. Muhammad, K.; Hamza, R.; Ahmad, J.; Lloret, J.; Wang, H.; Baik, S.W. Secure surveillance framework for IoT systems using probabilistic image encryption. *IEEE Trans. Ind. Inform.* **2018**, *14*, 3679–3689. [CrossRef]

44. Barthélemy, J.; Verstaevel, N.; Forehead, H.; Perez, P. Edge-Computing Video Analytics for Real-Time Traffic Monitoring in a Smart City. *Sensors* **2019**, *19*, 2048. [CrossRef]

45. Aqib, M.; Mehmood, R.; Alzahrani, A.; Katib, I.; Albeshri, A.; Altowaijri, S.M. Smarter Traffic Prediction Using Big Data, In-Memory Computing, Deep Learning and GPUs. *Sensors* **2019**, *19*, 2206. [CrossRef] [PubMed]

46. Xu, S.; Zou, S.; Han, Y.; Qu, Y. Study on the availability of 4T-APS as a video monitor and radiation detector in nuclear accidents. *Sustainability* **2018**, *10*, 2172. [CrossRef]

47. Plageras, A.P.; Psannis, K.E.; Stergiou, C.; Wang, H.; Gupta, B.B. Efficient IoT-based sensor BIG Data collection—Processing and analysis in smart buildings. *Futur. Gener. Comput. Syst.* **2018**, *82*, 349–357. [CrossRef]

48. Jha, S.; Dey, A.; Kumar, R.; Kumar-Solanki, V. A Novel Approach on Visual Question Answering by Parameter Prediction using Faster Region Based Convolutional Neural Network. *Int. J. Interact. Multimed. Artif. Intell.* **2019**, *5*, 30. [CrossRef]

49. Zhang, Q.; Wan, C.; Han, W. A modified faster region-based convolutional neural network approach for improved vehicle detection performance. *Multimed. Tools Appl.* **2019**, *78*, 29431–29446. [CrossRef]

50. Cho, S.; Baek, N.; Kim, M.; Koo, J.; Kim, J.; Park, K. Face Detection in Nighttime Images Using Visible-Light Camera Sensors with Two-Step Faster Region-Based Convolutional Neural Network. *Sensors* **2018**, *18*, 2995. [CrossRef]

51. Zhang, J.; Xing, W.; Xing, M.; Sun, G. Terahertz Image Detection with the Improved Faster Region-Based Convolutional Neural Network. *Sensors* **2018**, *18*, 2327. [CrossRef]

52. Liu, X.; Jiang, H.; Chen, J.; Chen, J.; Zhuang, S.; Miao, X. Insulator Detection in Aerial Images Based on Faster Regions with Convolutional Neural Network. In Proceedings of the IEEE International Conference on Control and Automation, ICCA, Anchorage, AK, USA, 12–15 June 2018; pp. 1082–1086.

53. Bakheet, S.; Al-Hamadi, A. A discriminative framework for action recognition using f-HOL features. *Information* **2016**, *7*, 68. [CrossRef]

54. Al-Gawwam, S.; Benaissa, M. Robust eye blink detection based on eye landmarks and Savitzky-Golay filtering. *Information* **2018**, *9*, 93. [CrossRef]

55. Krizhevsky, A.; Krizhevsky, A.; Sutskever, I.; Hinton, G.E. Imagenet classification with deep convolutional neural networks. *Adv. Neural Inf. Process. Syst.* **2012**. [CrossRef]

56. Simonyan, K.; Zisserman, A. Very Deep Convolutional Networks for Large-Scale Image Recognition. *arXiv* **2014**, arXiv:1409.1556.

57. Szegedy, C.; Liu, W.; Jia, Y.; Sermanet, P.; Reed, S.; Anguelov, D.; Erhan, D.; Vanhoucke, V.; Rabinovich, A. Going deeper with convolutions. In Proceedings of the IEEE Computer Society Conference on Computer Vision and Pattern Recognition, Boston, MA, USA, 7–12 June 2014; pp. 1–9.

58. He, K.; Zhang, X.; Ren, S.; Sun, J. Deep residual learning for image recognition. In Proceedings of the IEEE Computer Society Conference on Computer Vision and Pattern Recognition, Las Vegas, NV, USA, 26 June –1 July 2016; pp. 770–778.

59. Hou, R.; Chen, C.; Shah, M. Tube Convolutional Neural Network (T-CNN) for Action Detection in Videos. In Proceedings of the IEEE International Conference on Computer Vision, Venice, Italy, 22–29 October 2017; pp. 5823–5832.

60. Kalogeiton, V.; Weinzaepfel, P.; Ferrari, V.; Schmid, C. Action Tubelet Detector for Spatio-Temporal Action Localization. In Proceedings of the IEEE International Conference on Computer Vision, Venice, Italy, 22–29 October 2017; pp. 4415–4423.

61. Zolfaghari, M.; Oliveira, G.L.; Sedaghat, N.; Brox, T. Chained Multi-stream Networks Exploiting Pose, Motion, and Appearance for Action Classification and Detection. In Proceedings of the IEEE International Conference on Computer Vision, Venice, Italy, 22–29 October 2017; pp. 2923–2932.

62. Nvidia Corporation. Jetson Embedded Development Kit|NVIDIA. Available online: https://developer.nvidia.com/embedded-computing (accessed on 24 November 2019).

63. Abadi, M.; Agarwal, A.; Barham, P.; Brevdo, E.; Chen, Z.; Citro, C.; Corrado, G.S.; Davis, A.; Dean, J.; Devin, M.; et al. TensorFlow: Large-Scale Machine Learning on Heterogeneous Distributed Systems. *arXiv* **2016**, arXiv:1603.04467.

64. Nvidia Corporation. NVIDIA TensorRT|NVIDIA Developer. Available online: https://developer.nvidia.com/tensorrt (accessed on 24 November 2019).

65. Nvidia Corporation. NVIDIA DeepStream SDK|NVIDIA Developer. Available online: https://developer.nvidia.com/deepstream-sdk (accessed on 24 November 2019).

66. Fraga-Lamas, P.; Fernández-Caramés, T.M.; Suárez-Albela, M.; Castedo, L.; González-López, M. A Review on Internet of Things for Defense and Public Safety. *Sensors* **2016**, *16*, 1644. [CrossRef] [PubMed]

67. Gomez, C.A.; Shami, A.; Wang, X. Machine learning aided scheme for load balancing in dense IoT networks. *Sensors* **2018**, *18*, 3779. [CrossRef]

68. AMD Embedded Radeon^{TM}. Available online: https://www.amd.com/en/products/embedded-graphics (accessed on 24 November 2019).

MDPI

St. Alban-Anlage 66

4052 Basel

Switzerland

Tel. +41 61 683 77 34

Fax +41 61 302 89 18

www.mdpi.com

Information Editorial Office

E-mail: information@mdpi.com

www.mdpi.com/journal/information